MIX
Papier aus verantwortungsvollen Quellen
Paper from responsible sources
FSC® C105338

Ashutosh Mokate

VIBRATION AND NOISE REDUCTION IN PLANETARY GEAR TRAIN BY PHASING

Anchor Academic Publishing

Mokate, Ashutosh: VIBRATION AND NOISE REDUCTION IN PLANETARY GEAR TRAIN
BY PHASING, Hamburg, Anchor Academic Publishing 2016

Buch-ISBN: 978-3-96067-018-6
PDF-eBook-ISBN: 978-3-96067-518-1
Druck/Herstellung: Anchor Academic Publishing, Hamburg, 2016

Bibliografische Information der Deutschen Nationalbibliothek:
Die Deutsche Nationalbibliothek verzeichnet diese Publikation in der Deutschen
Nationalbibliografie; detaillierte bibliografische Daten sind im Internet über
http://dnb.d-nb.de abrufbar.

Bibliographical Information of the German National Library:
The German National Library lists this publication in the German National Bibliography.
Detailed bibliographic data can be found at: http://dnb.d-nb.de

All rights reserved. This publication may not be reproduced, stored in a retrieval system
or transmitted, in any form or by any means, electronic, mechanical, photocopying,
recording or otherwise, without the prior permission of the publishers.

Das Werk einschließlich aller seiner Teile ist urheberrechtlich geschützt. Jede Verwertung
außerhalb der Grenzen des Urheberrechtsgesetzes ist ohne Zustimmung des Verlages
unzulässig und strafbar. Dies gilt insbesondere für Vervielfältigungen, Übersetzungen,
Mikroverfilmungen und die Einspeicherung und Bearbeitung in elektronischen Systemen.

Die Wiedergabe von Gebrauchsnamen, Handelsnamen, Warenbezeichnungen usw. in
diesem Werk berechtigt auch ohne besondere Kennzeichnung nicht zu der Annahme,
dass solche Namen im Sinne der Warenzeichen- und Markenschutz-Gesetzgebung als frei
zu betrachten wären und daher von jedermann benutzt werden dürften.

Die Informationen in diesem Werk wurden mit Sorgfalt erarbeitet. Dennoch können
Fehler nicht vollständig ausgeschlossen werden und die Diplomica Verlag GmbH, die
Autoren oder Übersetzer übernehmen keine juristische Verantwortung oder irgendeine
Haftung für evtl. verbliebene fehlerhafte Angaben und deren Folgen.

Alle Rechte vorbehalten

© Anchor Academic Publishing, Imprint der Diplomica Verlag GmbH
Hermannstal 119k, 22119 Hamburg
http://www.diplomica-verlag.de, Hamburg 2016
Printed in Germany

TABLE OF CONTENTS

CHAPTER 1 – INTRODUCTION 1
    1.1 Problem Statement 4
    1.2 Objectives 5
    1.3 Scope 6
    1.4 Methodology 7
    1.4.1 Method of Phasing Gears 7
    1.5 Organization of the Study 9

CHAPTER 2 – LITERATURE REVIEW 10

CHAPTER 3 – MATERIAL SELECTION FOR GEAR & THEIR PROTECTION ... 28
    3.1 Selection of Plastic Materials 28
    3.2 Why Plastic? 29
    3.3 Mechanical Properties of Various Plastics 30
    3.4 Properties of Nylon-6 30
    3.5 Applications of Nylon-6 30

CHAPTER 4 – EXPERIMENTAL SET-UP AND MEASUREMENTS 31
    4.1 Motor Selection 32
    4.2 Lovejoy Coupling 33
    4.3 Selection of Gear Box 34
    4.4 Coupler Shaft 34
    4.5 Selection of Gear Box Gear Pair-1 & II (Phase II) 34
    4.6 Gear Pair-3 (Phase II) 34
    4.7 Gear Pair-4 (Phase II) 34
    4.8 Sound Level Meter 34
    4.9 FFT Analyzer 35
    4.10 Phasing Arrangement 37
    4.11 Noise Measurement in PGT 41

CHAPTER 5 – RESULTS AND DISCUSSION 43
5.1 Noise Measurement in single PGT, Without Phasing & With Phasing arrangement 43
5.2 Vibration measurement in single planetary gear trains 47
5.3 Vibration measurement in PGT by without & with phasing of gear pair 50
5.4 Acceleration Spectrums 57
5.5 Displacement spectrums 73
5.6 Velocity Spectrums 89
CHAPTER 6 – CONCLUSION 105
FUTURE SCOPE 106
REFERENCES 107

CHAPTER 1

INTRODUCTION

Gears are essential parts of many precision power transmitting machine such as an automobile. The major functions of a gearbox are to transform speed and torque in a given ratio and to change the axis of rotation. Planetary gears yield several advantages over conventional parallel shaft gear systems. They produce high speed reductions in compact spaces, greater load sharing, higher torque to weight ratio, diminished bearing loads, and reduced noise and vibration.

They are used in automobiles, helicopters, aircraft engines, heavy machinery, and a variety of other applications. Despite their advantages, the noise induced by the vibration of planetary gear systems remains a key concern. Planetary gears have received considerably less research attention than single mesh gear pairs. There is a particular scarcity of analysis of two planetary gear systems and their dynamic response. This paper focus on the study of two PGTs with different phasing (angular positions) while keeping every individual set unchanged. (01)

Planetary gear systems are used to perform speed reduction due to several advantages over conventional parallel shaft gear systems. Planetary gears are also used to obtain high power density, large reduction in small volume, pure torsion reactions, and multiple shafting. Another advantage of the planetary gearbox arrangement is load distribution. Because the load being transmitted is shared between multiple planets, torque capability is greatly increased. If the number of planets in the system are more the ability of load shearing is greater and the higher the torque density. The planetary gearbox arrangement also creates greater stability due to the even distribution of mass and increased rotational stiffness. Despite their advantages the noise induced by vibrations of planetary is concern, particularly in automotive industry where the vehicle interior noise is a key quality metric.

Noise and vibration generated in gears is mainly due to the transmission error, this is the difference between the position of the driven gear without torque and manufacturing errors, and the actual position including all those effects. Reducing the amplitude of the transmission error is possible by selecting suitable profile modifications. Extensive research work has been carried out by many researchers on the analysis of errors, dynamic response, and noise and vibration reduction in single planetary gears. They are using the various methods of reducing the vibration and noise in planetary gear by changing the number of teeth and by using the analytical as well as FEM for reducing the vibration and noise.

Planetary gears are very popular due to their advantages such as high power density, compactness, and multiple and large compact gear ratios and load sharing among planets. Gearing arrangement is comprised of four different elements that produce a wide range of speed ratios in compact layout. These elements are, (1) Sun gear, an externally toothed ring gear co-axial with the gear train (2) Annulus, an internally toothed ring gear coaxial with the gear train (3) Planets, externally toothed gears which mesh with the sun and annulus, and (4) Planet Carrier, a support structure for planets, co-axial with the train. Planetary gear system as shown in Figure 1 is typically used to perform speed reduction due to several advantages over conventional parallel shaft gear systems. Planetary gears are also used to obtain high power density, large reduction in small volume, pure torsional reactions and multiple shafting. Another advantage of the planetary gearbox arrangement is load distribution. If the number of planets in the system are more the ability of load shearing is greater and the higher the torque density. The planetary gearbox arrangement also creates greater stability due to the even distribution of mass and increased rotational stiffness.

In recent years, enhancement of interior quietness in passenger cars, Automobiles is an important factor for influencing occupant comfort. Planetary gear sets are essential components of automatic transmissions because of their compact size and wide gear ratio range. They produce high speed reductions in compact spaces, greater load sharing, higher torque to weight ratio, diminished bearing loads and reduced noise and vibration. A Despite their advantage, the noise induced by the vibration of planetary gear systems remains a key concern. Planetary gears have received considerably less research attention than single mesh gear pairs. This paper focus on the study of two PGTs with different phasing (angular positions) while keeping every individual set unchanged.

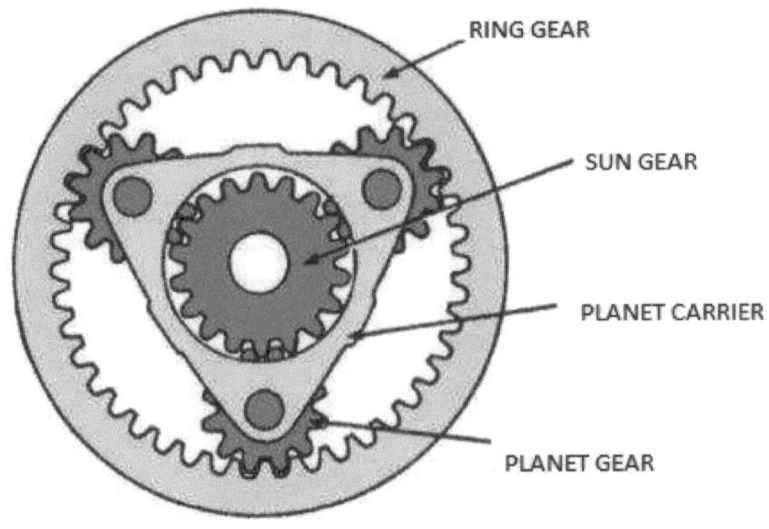

Figure 1: Basic layout of planetary gear box

This figure shows that the basic layout planetary gear train in which there is one Sun gear, Three Planet gear and one ring gear. They can produce the high speed reduction in compact space and having greater load shearing capacity & high torque to weight ratio.

1.1 Problem Statement

The vibration occurs due to the improper meshing between the gear and continuous wear of gear that occurs because of continuous machining operation & Defects in manufacturing. Due to the vibrations in the system noise is generated and also it impacts on the system performance.

1.2 Objectives

The main objective of project is to reduce the vibration and noise in the planetary gear train.

The objectives are as follows –

1. Development of experimental setup for reduce the Vibration and Noise in planetary gear train.
2. Measure the vibration with single PGT arrangement.
3. Measure the vibration & Noise in PGT with phasing arrangement.
4. Measure the vibration & Noise in PGT without phasing arrangement.
5. Comparison of results between phasing and without phasing arrangement.

1.3 Scope

Vibration and noise generated in the planetary gear train is reduced by phasing arrangement between two planetary gear sets. The phasing angle which is obtained from the number of teeth is provided between two gear sets for measuring the vibration and noise in gear train. The phasing angle to be used in planetary gear train should be as per calculated from number of teeth.

1.4 Methodology

After developing the experimental setup the vibrations and noise are measured by using FFT analyzer and sound measuring instrument. In the first part the vibration and noise are measured by without phasing arrangement of planetary gears after that the vibration and noise are measured by phasing arrangement between the planetary gear pair. After taking the results of both compare the results of noise and vibration with and without phasing arrangement.

1.4.1 Method of Phasing Gears

To control the vibrations in tooth gearings effectively, one should have an adequate knowledge of the physical nature of what causes vibrations in planetary gear pair with imprecise and deformed teeth. Vibrations in gearing is caused by an internal excitations, as it occurs at the contact of two compressed elastic bodies (teeth) during their relative motion and acts on both bodies with the same intensity but in opposite directions.

Because the variation of tooth mesh stiffness during meshing as a principal source of internal excitation force and vibration, modifications of the optimal tooth shape and contact ratio (CR) have been studied as ways of reducing the variation in mesh stiffness. Major variations in stiffness are caused by changes in meshing pair numbers, usually in the range 1.0-2.0 for normal spur gears. It is impossible to avoid this variation due to the integer numbers of gear teeth.

If another meshed and phased gear pair is added to reverse the stiffness functions of the two pairs, these phasing gears will complement the primary gears and reduce the mesh stiffness variation. The phasing gear pair is made up of two gears half the width and half the pitch phasing of the primary gears. The conceptual model of phasing gears is shown in Figure 2.

Figure 2: Conceptual model of phasing of gear pair

In this gear pair the angle is provided between the two teeth is depend on the number of teeth, is the number of teeth varies the angle between the two gears will be change. This figure shows the inclination of one gear pair by keeping another gear fixed.

1.5 Organization of the Study

Organization of this Study includes:

1) 1st chapter is related to the introduction of planetary gear in which the advantages of PGT are discussed and collecting the problems found in gear pair and also includes the Problem statement, objectives, Scope, Methodology.

2) 2nd chapter is related to the Literature Survey.

3) 3rd chapter is related to the Experimental Validation in this chapter the information of experimental setup is given.

4) 4th chapter is related to Results and discussion.

5) 5th chapter is related to the conclusion and future scope.

CHAPTER 2

LITERATURE REVIEW

R.G.Parkar et al. (2000) studied the dynamic response of planetary gear was of the fundamental importance in helicopters, automotive transmissions, aircraft engines, and a variety of industrial machinery. The complex, dynamic forces at the sun, planet and ring planet meshes were the source of the vibration. Modeling of the dynamic tooth forces remains an important issue that has not been resolved even for single-mesh gear pairs. The multiple meshes of planetary gears further complicate the dynamic modeling. Consequently, dynamic analyses of planetary gears were less developed than for other single-mesh gear configurations. In particular, experimental frication of the existing analytical models was especially limited.

As a result, design options to minimize noise and loads in planetary gears have developed empirically without strong analytical or experimental foundation. Their design strategies include tooth-shape medications, gear geometry adjustments (pitch, contact ratio, etc.), reduction of manufacturing tolerances, use of sun, ring, or carrier components, and vibration isolation concepts. A particular strategy was the use of planet phasing, where the planet configuration and tooth numbers were chosen such that self-equilibration of the mesh forces reduces the net forces and torques on the sun, ring, and carrier, thereby reducing vibration. Figure 3 shows the schematic layout of planetary gear in which the mesh force was acting on the planet gear. (5)

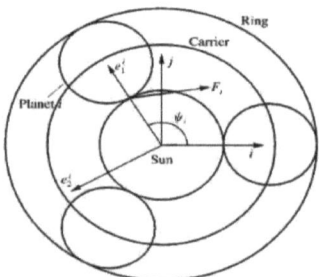

Figure 3: Planetary gear schematic. Fi denotes the mesh force at the *i* th sun planet mesh

C.Gill-Jeong et al. (2010) studied that gearing assembly is one of the major vibration source in power transmission system especially used in automotive, aerospace, marine and industrial applications. Their study presents a novel means of reducing gear vibrations using a simple 1:1 ratio spur gear pair using a method of phasing. Variation in the gear mesh stiffness over a mesh cycle which depends on the number of pairs of teeth in contact was one of the principal causes of vibrations and instabilities and has a strong influence on the overall dynamics of the geared system.

Their method was based on reducing the variation in gear mesh stiffness by adding another pair of gears with phasing. Because of added phasing gear, the numbers of pairs of teeth in contacts were increased which reduces the variation in mesh stiffness. A simple spur gear model with rectangular-wave-type mesh stiffness was assumed and mesh stiffness variation was obtained numerically using MATLAB 7.5 software and was comparable in both cases i.e. normal and phasing gears.

Their numerical result of analysis shows the reduction in mesh stiffness variation and the possibility of reduction in vibration in simple spur gear pair using the proposed method. (6)

Majid Mehrabi, Dr. V. P. Singh (2013) works on planetary gearboxes were usually used in a wide variety of machinery such as automobiles, helicopters and aircraft engines .Their numerous advantages are high speed reductions in compact spaces, high torque/weight ratio, greater load sharing, diminished bearing loads and reduced noise and vibration. A typical simple planetary gear set consists of a sun gear, a ring gear and a number of identical planet gears (typically 3–6) meshing both with the sun and ring gears.

They were well known for their symmetrical structure which allows an equal share of the total external torque applied between the planetary gears, the sun and the ring. However non stationary conditions of system such as overload conditions, torque fluctuation may affect the dynamic behavior of a planetary gear transmission.

The inequality of the load distribution however arises on each planet gear because of random errors of manufacture, assembly and operating conditions. Their results in noise and vibration which were key concerns in their applications and drop in efficiency of planetary gear system. In some helicopters planetary gear vibration was the primary source of cabin noise that can exceed 100dB. Before 1990, their literature on analytical planetary Gear dynamics was scare. They studied the Eigen value problem for a thirteen degree of freedom system and identified the natural frequencies and vibration modes. (21)

A. Kahraman, R. Singh (1990) Discussed on the dynamic response of gears which is related to the noise generation and dynamic loads. Prior studies have yielded a vast literature on their topic and, in particular, a remarkable variety of mathematical models as discussed in reference. More recent studies were cited in their comprehensive bibliography in reference. Most models were use a discrete (lumped parameter) representation involving rigid gear components and combinations of discrete elastic and dissipative elements to represent the meshing teeth and support/bearing stiffness. Such models have varying complexity in their treatment of the tooth mesh, shaft, bearing, and housing modeling. In essence, the required analytical modelling to capture the complex gear dynamic response has not been established.

Even when attention was restricted to modelling the tooth mesh, a variety of possible representations exist, and the optimal treatment of time-varying mesh stiffness, contact loss, use of static transmission error as a dynamic input, frictional effects, etc., remains unsettled. Their study analytically investigated the dynamics of a spur gear pair for which comprehensive experimental data exist.

Their tooth mesh was the most complex aspect in gear dynamics, and the gear system in that work was selected to isolate tooth mesh effects. Their primary analytical tool was infinite elements/contact mechanics (FE/CM) formulation that offers significant advantages in its representation of the crucial tooth contact. Their purpose was to further expose the basic non-linear and time-varying phenomena at play in the tooth mesh, demonstrate the modelling fidelity and advantages of the FEM method used, and compare the ability of two s.d.o.f models to represent the experimentally observed phenomena.

They studied the gear pair that used in a series of experiments by Kahraman and Blankenship. Tests on their system were initially reported in reference, where the details of the system were given. Their test stand was designed to isolate the impact of tooth mesh interactions on the dynamic response and exclude complications from the shafts, bearings, and housing. In particular, the bearing and shaft configuration was such that the support structure was nearly rigid and the response was purely gear rotation. The test gears were dynamically isolated from the slave gears in the back-to-back configuration. Despite their reduction to the simplest case of s.d.o.f response, measurements of dynamic transmission error show distinct, repeatable, non-linear, time varying system response in the form of classical jump phenomena, sub- and super harmonic resonances, parametric instabilities, and even apparently chaotic response. The non-linear tooth mesh forces causing these complex behaviors were what they seek to model in their study.

A primary motivation was to establish the ability of the unique FE/CM formulation to capture complex gear mesh forces in dynamics simulations. Similar analysis tools with their advantages presented in what follows were not known to the authors. Conventional "infinite element analysis, and even the currently available commercial software, require prohibitively redefined meshes to represent the tooth contact and precise tooth surface description needed for gear mechanics, particularly when one seeks to go beyond static analyses to dynamic response analyses. Their subject gear system was selected to validate the FE/CM approach as a research tool because the complex, non-linear behavior was suitably demanding benchmark, and carefully conducted, high-quality experiments exist.

The finite element formulation was unique in its combination of detailed contact modelling between the elastic teeth with a combined surface integral finite element solution especially capture tooth deformations and loads with a relatively coarse mesh. Details were available in the references and a short description of the surface integral finite element solution was given in reference. The contact analysis was briefly described there. They mesh for the gear pair in their study was shown. Each of the gears undergoes large rotation according to a prescribed, kinematic trajectory. In that two-gear case, there trajectory was that of conjugate action of the gears at specified operating speed.

The elastic gear motions that superpose on their prescribed trajectory are small. If the infinite element displacement vector x fit for a particular gear i was measured with respect to a reference frame that follows that known trajectory, then it was possible to represent its behavior by a linear system of equations. The contact analysis was briefly described there. The mesh for the gear pair in their study was shown. Each of the gears undergoes large rotation according to a prescribed, kinematic trajectory.

In that two-gear case, the trajectory was that of conjugate action of the gears at specified operating speed. (13)

Figure 4 shows the single degree of freedom system having equal in radius and moment of inertia of two different gears which was connected by spring and damper in between two gears which was shown in figure 4.

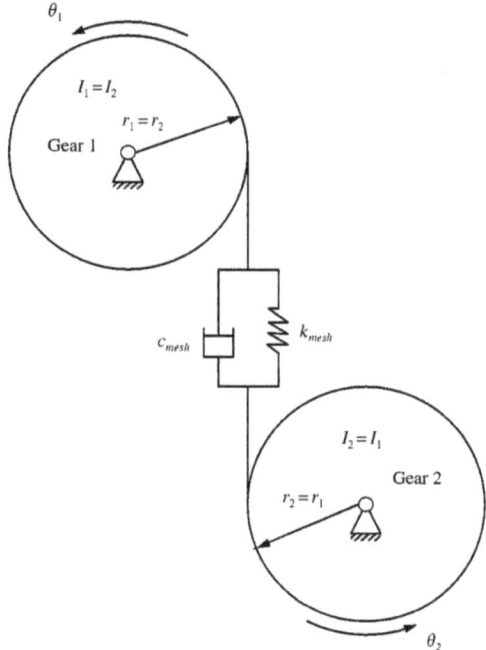

Figure 4: Single- degree-of-freedom modelling of the two gear system

R.G.Parker, X. Wu (2002) they use a planetary gear at particular strategy to reduce vibration by using planet phasing, where the planet configuration and tooth numbers were chosen such that self-equilibration of the mesh forces reduces the net forces and torques on the sun, ring, and carrier, thereby reducing vibration. Their idea was proposed by Schlegel and Mard where experimental measurements on a spur gear system demonstrated a noise reduction of 11 dB Seager gave a more detailed analysis using a static transmission error model of the dynamic excitation. Palmer and Fuehrer also demonstrate the effectiveness of

Planet phasing and support their arguments with limited experiments. Kahraman and Kahraman and Blankenship studied the use of planet phasing in the context of helical planetary systems.

Their work use static transmission error to represent the dynamic excitation in a lumped parameter dynamic model. Their work examines the analytical basis for planet phasing in spur planetary systems. The results were developed in terms of the physical mesh forces and were not tied to any lumped parameter model. In fact, they make no attempt to characterize the factors affecting the mesh forces or quantify their magnitudes. The fundamental issue was that the inherent symmetries of planetary gears imply distinct relationships between the dynamic forces at the individual meshes. Their approach was more appealing to physical intuition and makes no supposition about the use of static transmission error to model the dynamic excitation.

The symmetries lead naturally to specific conclusions for the suppression of particular harmonics of mesh frequency in the net forces and torques on the sun, ring, and carrier. Simple rules with clear design application emerge to suppress expected resonances that occur when the mesh frequency M or one of its harmonics coincides with a system natural frequency. Systems with equal planet spacing were examined in detail, although the methods adapt easily to systems with unequal spaced, diametrically opposed planets, which was the basis for the work presented there. The unique structure of planetary gear vibration modes was essential in what follows. Planetary gears with equal sun planet mesh stiffness at each mesh, equal ring planet mesh stiffness at each mesh, and equal planet inertia properties have exactly three types of modes for systems with equally spaced or diametrically opposed planets: rotational, translational, and planet modes. When the elastic deformation of the ring gear was included that same mode types persist and a fourth category of purely ring modes was added. In order to maximize the power density and improve load sharing among the planets, planetary gears in numerous industries were designed to have thin ring gears, and this leads to elastic deflection of the ring gear.

Most works model the ring gear as a rigid body. Wu and Parker established an elastic discrete model that includes planetary gear discrete degrees of freedom (rotational and translational) and ring gear elastic deflection. That model was adopted in, which was the basis for the material in that paper.

Gear vibration was driven by changing mesh stiffness as the number of teeth in contact changes. That was often modelled as time varying parametric excitation, which was closely related to modelling based on static transmission error as an external "right hand side" driving force. Parametric instabilities were demonstrated clearly in experiments on a spur gear pair Measurements showed large resonant vibration when the mesh frequency was twice the natural frequency, which was a classical signature of parametric instability. The amplitudes of vibration became sufficiently large as a result of parametric instability that nonlinear phenomena such as tooth contact loss, period- doubling, and chaos were also observed. Mathematical modelling with parametric excitation from varying mesh stiffness as the driving excitation source agreed well with the experiments, including for speed ranges where mesh frequency nearly equals a natural frequency, which was the most widely studied gear resonance condition Parametric instability in single - pair gears has been investigated in . Only a few studies exist on parametric instabilities of multiple mesh gear systems. Tordion and Gauvin and Benton and Seireg analyzed the instabilities of two stage gear systems but with contradictory conclusions. That was clarified by Lin and Parker, who derived formulae that allow designers to suppress particular instabilities by choice of contact ratios and mesh phasing. Liu and Parker analytically investigated the nonlinear resonant vibration of idler gears parametrically excited by mesh stiffness variation.

The impact of mesh stiffness variation on tooth loads and load sharing in planetary gears was studied by August and Kasuba and Velex and Flamand. They numerically computed the dynamic response of planetary gears with three sequentially phased meshes and found the impact of mesh stiffness variations on dynamic response was significant. Lin and Parker analytically investigated the parametric instability of planetary gears using a purely rotational model, and Bahk and Parker extend that to examine the nonlinear dynamics. All of their works adopt a rigid ring model. Their work which was from, examines planetary gear parametric instability using a model that includes the translational vibration of all components and the elastic deformation of the ring gear. With the modal expressions of the elastic-discrete model from, the instability boundaries were obtained as simple expressions. They show that many modes cannot interact to create combination instabilities, and general instability existence rules were obtained for equally spaced planets. By adjusting the tooth numbers, contact ratios, and mesh phases, one can minimize or completely suppress much potential instability. (11)

Kahraman C. Yuksel (2004) studied on Planetary gear sets, also known as epicyclic gear drives, were commonly used in a large number of automotive, aerospace and industrial applications. They possess numerous advantages over parallel-axis gear trains including compactness of design, availability of multiple speed reduction ratios, and less demanding bearing requirements. Most common examples of planetary gear sets can be found in automatic transmissions, gas turbines, jet engines, and helicopter drive trains. A typical simple planetary gear set consists of a sun gear, a ring gear and a number of identical planet gears (typically 3–6) meshing both with the sun and ring gears.

A common carrier holds the planets in place. Dynamic analysis of planetary gears was essential for eliminating noise and vibration problems of the products they are used in. The dynamic forces at the sun- planet and ring-planet meshes were the main sources of such problems. Although planetary gear sets have generally more favorable noise and vibration characteristics compared to parallel-axis gear systems, planetary gear set noise still remains to be a major problem. The dynamic gear mesh loads that were much larger than the static loads were transmitted to the supporting structures, in most cases, increasing gear noise. Larger dynamic loads also shorten the fatigue life of the components of the planetary gear set including gears and bearings.

Surface wear was considered one of the major failure modes in gear systems. In case of planetary gear sets, experimental data has shown that especially the sun gear meshes might experience significant surface wear when run under typical operating conditions. While wear was a function of a large number of parameters, sliding distance and contact pressure were shown to be most significant parameters influencing gear wear. Wear of tooth profiles results in a unique surface geometry that alters the gear mesh excitations in the form of kinematic motion errors, enhancing the dynamic effects. Modelling of planetary gear set dynamics received significant attention for the last 30 years. A number of studies proposed lumped-parameter models to predict free and forced vibration characteristics of planetary gear sets. Their models assumed rigid gear wheels, connected to each other by springs representing the flexibility of the meshing teeth. In their studies, nonlinear effects due to gear backlash and time-varying parameters due to gear mesh stiffness fluctuations were neglected. The corresponding Eigen value solution of the linear equations of motion resulted in natural modes. Modal summation technique was typically used to find the forced response due to external gear mesh displacement excitations defined to represent motion transmission errors. That lumped-parameter model varies in degrees of freedom included, from purely torsional models to two or three-dimensional transverse-torsional models.

While that model served well in describing the dynamic behavior of planetary gear sets qualitatively, they lacked certain critical features. First, the gear mesh models were quite simplistic with a critical assumption that complex gear mesh contact interaction can be represented by a simple model formed by a linear spring and a damper. That models demand that the values of the gear mesh stiffness and damping, as well as the kinematic motion transmission error excitation, must be known in advance.

It was also assumed that these parameter values determined quasi-statically remain unchanged under dynamic conditions. In addition, gear rim deflections and Hertzian contact deformations were also neglected. Another group of recent models used more sophisticated finite element-based gear contact mechanics models. There computational models address all of the shortcomings of the lumped-parameter models since the gear mesh conditions were modelled as individual nonlinear contact problems. The need for externally defined gear mesh parameters was eliminated with their models. In addition, rim deflection and spline support conditions were modelled accurately.

Their models were also capable of including the influence of the tooth profile variations in the form of intentional profile modifications, manufacturing errors or wear on the dynamic behaviour of the system. 696 C. Yuksel, A. Kahraman the study of wear of gear contact was becoming one of the emerging areas in gear technology. A number of recent gear wear modelling efforts form a solid foundation for more accurate, larger system analyses. All of these models use Archard's wear model in conjunction with a gear contact model and relative sliding calculations.

Their studies focused on prediction of wear of either spur or helical gear pairs in a parallel-axis configuration. The tooth contact pressures were computed in that models using either simplified Hertzian contact or boundary element formulations under quasi-static conditions. Sliding distance calculations were carried out kinematically by using the involutes geometry and Archard's wear model was used with an empirical wear coefficient to compute the surface wear depth distribution.

A number of studies investigated the influence of wear on gear dynamics response. Among them, Kuang and Lin simulated the tooth profile wear process, and predicted the variations of the dynamic loads and the corresponding frequency spectra as a function of wear for a single spur gear pair. Wojnarowski and Onishchenko performed analytical and experimental investigations of the influence of the tooth deformation and wear on spur gear dynamics. They stated that the change in the profiles of the teeth due to wear must be taken into account when dealing with the durability of the gear transmissions as well.

Their previous models considered surface wear effects for only a single spur gear pair, voiding multi-mesh gear systems such as the planetary gear sets. They focused on only external gears and used lumped-parameter dynamic models excluding nonlinear and time-varying effects. (4)

Yichao Guo, (2011) discussed on planetary gears were widely used in all kinds of transmission systems, such as wind turbines, aircraft engines, automobiles, and machine tools and they were classified into two categories: simple and compound planetary gear. Simple planetary gears have one sun, one ring, one carrier, and one planet set (i.e., single stage). There was only one planet in each planet train

Compound planetary gears involve one or more of the following three types of structures: meshed-planet (there are at least two more planets in mesh with each other in each planet train), stepped-planet (there exists a shaft connection between two planets in each planet train), and multi-stage structures (that system contains two or more planet sets). Compared to simple planetary gears, compound planetary gears have the advantages of larger reduction ratio, higher torque-to-weight ratio, and more flexible configurations.

In spite of their advantages, vibration remains a major concern in planetary gear applications. Vibration creates undesirable noise, reduces fatigue life of the whole system, and decreases durability and reliability. Vibration reduction was key to the applications of compound planetary gears. They require analytical study on compound planetary gear dynamics to provide fundamental understanding of the dynamics and guide vibration reduction.

Most research on gear dynamics focuses on single gear pairs or multi-mesh gear systems. Recently, considerable progress has been made in the modeling and analysis of simple planetary gear. Studies on compound planetary gears, however, were limited. Many fundamental analyses that were proved to be essential in other systems and studies have not been performed, including the purely rotational system modeling and the associated modal properties, the impact of system parameter changes on natural frequencies and vibration modes (Eigen sensitivity analysis), the natural frequency veering and crossing patterns, the classification of mesh phase relations, the suppression of selected dynamic responses through mesh phasing, and the parametric instability caused by mesh stiffness variations were not performed. (11) Their study aims at their research gaps and the main objectives were:

- To develop a purely rotational model for general compound planetary gears that can clarify the confusion in previous rotational planetary gear models and analytically prove the associated modal properties.
- To perform an Eigen sensitivity analysis based on Kiracofe and Parker's rotational-translational model and derives the Eigen sensitivities in compact formulae.
- To inspect the natural frequency veering/crossing phenomena and identify any patterns or general rules.
- To find a way to analytically describe and calculate all the relative mesh phases in a compound planetary gear.
- To investigate the existence of mesh phasing rules for deferent compound planetary Gear models that can suppress vibration.
- To study the parametric instability caused by mesh stiffness variations and to Analytically determine the boundaries for instability regions.
- To examine the back-side mesh stiffness and to quantify the impact of backlash on the back-side mesh stiffness.

A. Palermo et al. (2010) studied on gears which were extensively employed in mechanical systems since they allow the transfer of motion in a wide range of working conditions, with a variety of gear ratios, and at reasonable production costs. The gear meshing was a complex process because it involves moving and multiple contact points, variable load sharing on the meshing teeth, contact mechanics (which is nonlinear), and all of them from a dynamic standpoint. Furthermore tooth micro geometry, manufacturing imperfections and assembly errors have relevant effects on the behavior of gear systems

And cannot be ignored. That complexity has to be faced in the design phase, which must address both endurance and noise requirements. For that reasons, a considerable amount of research works on gear dynamics was available in literature, but still many aspects remain unresolved. Moreover, today's markets were highly competitive and therefore reaching valid solutions in shorter timeframes represents a clear advantage. In that context, numerical models and simulations allow to achieve solutions to improve the dynamic behavior of gear systems, and to limit the testing phase saving time and money. That explains why efforts continue to be spent in the gear dynamics research field, with applications especially in helicopter, wind turbine and automotive industries.

Gear dynamics were mentioned because, at the occurrence, the proposed methodology enables to evaluate the dynamic meshing loads, which in transients can be several times higher than the static ones. From that perspective, the proposed technique will also allow the simulation of load sharing in planetary gear trains, which was currently a major issue for that kind of transmissions. Coming back to gear noise purposes, the proposed methodology takes into account the dynamic transmission error (DTE), which was defined for a gear pair as the dynamic relative displacement between meshing teeth.

The transmission error was widely regarded as one of the main causes of gear noise. According to Munro, the transmission error was defined first by Harris in 1958, which started its analytical investigation. Nevertheless references prove that their concept was already applied before, but using an empirical approach. The two main factors affecting the TE were the mesh stiffness, which accounts for tooth flexibility and number of meshing tooth pairs, and the tooth micro geometry, in terms of intentional modifications and manufacturing errors. Variations in the TE, during the gear meshing, trigger vibrations and then airborne noise. Their analysis will be focused on the case of involutes parallel spur and helical gears, which were the most common ones.

Several methodologies were available in literature to simulate and estimate meshing vibrations, using analytical lumped parameter, Finite Element (FE), and multi body approaches. The analytical model proposed by Umezawa, later corrected by Cai to consider the influence on tooth stiffness of the gear tooth number, describes the gear meshing with a single degree of freedom (SDOF) system aligned along the line of action. Assuming a time-varying function for the mesh stiffness, defined within one mesh period (or, a dimensionally, within one mesh cycle), and the damping, the equation of motion can be solved. Their models allow considering the effects of tooth micro geometry, assembly and manufacturing errors, lumping them on the line of action with a displacement-driven excitation for the SDOF system.

With that assumptions it was impossible to consider the three-dimensionality of the contact problem, and the quality of the results that can be obtained depends on how realistic were the mesh stiffness function for the given gear pair and the displacement excitation. Different analytical models improved the accuracy of the results using a FE model to take into account shaft deflections and three-dimensional geometry, but the lumped parameters description was still suitable to analyze simple cases. Full FE models allow more accurate representations of gear systems and avoid a-priori assumptions on the TE, but since tooth contact happens in a very small area, and it spans the teeth from root to tip, highly refined mesh or contact detection followed by re meshing was needed along the whole tooth face. That causes high computational costs to run a simulation. Moreover, it was also an issue to correctly describe the tooth three-dimensional micro geometry. In the FE field, an interesting technique was proposed by Parker et al.

They use a semi-analytical finite element formulation specifically devised for contact problems. The tooth was divided in a contact zone (extending beneath the tooth surface) and a FE zone, separated by a matching interface. The contact zone was analytically solved by means of the Boussinesq's solution. That solution was evaluated at the matching FE nodes and the obtained nodal parameters were used to solve the remaining

FE part. However, FEM techniques were not able to consider with good computational efficiencies nonlinear entities such as gear lashes, assembly clearances, bearings, clutches, and other nonlinear effects which arise from large angle rotations. Such nonlinear effects require time-domain integration, which was typical of the multi body environment. Moreover, to assess the noise and vibration performances of a geared system it was usually desirable to test a wide range of working conditions (e.g., torques, regime run-up,). With such demands, the use of large scale finite element models in time domain becomes computationally expensive and maybe impractical.

The technique proposed in that paper was an improvement of the Static Transmission Error method described by Morgan et al. The basic idea of the method was to let specialized, thus highly efficient, software for gear contact analysis (abbreviated GCAS from now on) execute the calculation of the static mesh stiffness, which can then be used in the dynamic multi body simulation. The gear meshing, in the multi body simulation, was in fact governed by the dynamic equilibrium of the contact forces applied to the gears, rather than the ideal kinematic contact ratio. Considering a single spur or helical gear pair described by the standard gear parameters, the static mesh stiffness can be calculated using GCAS which enables to take into account three-dimensional teeth micro geometric modifications and manufacturing errors, teeth global and contact stiffness, shaft deflections and assembly misalignments. That mesh stiffness was obtained, for one static working condition, as a function of the position along the mesh cycle.

Once the static mesh stiffness was imported in the multi body software, the contact forces were calculated and applied to the gears by a user-defined force element which reads the instantaneous value of the mesh stiffness based on the actual position along the mesh cycle. In this way, the meshing complexity is captured avoiding the high computational cost related to the full-scale model, since the multi body gears and shafts models were rigid (while the bearings are compliant). The improvement brought by the current work was the capability to import and use the static mesh stiffness as a function of the instantaneous values of torque. In that paper, first the multi body model adopted for the gear system is described, and then the Static Transmission Error (STE) and the Dynamic Transmission Error (DTE) were defined. Subsequently, the static mesh stiffness sensitivity to the main assembly errors was evaluated, in order to identify the most influent ones. A description of the new technique to consider the variable torque follows. Finally, the obtained results were discussed and compared to the ones obtained with the previous technique and the static GCAS values. (15)

A Al-Shyyab et al. (2009) discussed on planetary gear sets which were widely used in many applications including automotive transmission, rotorcraft, wind and gas turbine gearboxes, as well as other marine and industrial power transmission systems. Planetary gear trains have many advantages over fixed center counter-shaft gear systems. The flow of power via multiple-gear meshes increases the power density, helping to reduce the overall size of the gearbox. The axisymmetric orientation of the planet gears reduces the radial bearings loads and in many cases allowing its central members (sun gear, ring gear, or the planet carrier) to float radially. That was reduces the effect of gear and carrier

Manufacturing errors on planet load sharing. Finally, the ability of multi-stage planetary sets in providing multiple speed reduction (gear) ratios has been the main reason for their extensive use for automatic transmission applications. Compound planetary trains were obtained from number of single-stage planetary gear sets whose central members were connected according to a given power flow configuration. Input, output, and fixed (stationary) member assignments were made to certain central members to achieve a given gear ratio.

Most of the planetary gear train dynamic models were limited to single-stage planetary gear sets. Early models were of linear time-invariant type (no backlash and constant mesh stiffness) where the Eigen solutions and model summation techniques were used to predict the natural modes and the forced response. That model were extended to study the neutralization or cancellation of excitations at each gear mesh through proper phasing of the gear meshing by specifying the planet position angles and numbers of teeth of gears. A schematic of a three-stage segment of a stage compound gear train was considered. A particular stage compresses three central elements, the sun gear, the ring gear and carrier as well as number of planets. The planets of each stage were free to rotate with respect to their common carrier, while the central elements of any stage were candidates for being an input, output, or reaction member. Multiple stages can be connected in different ways via a proper coupling central member of stage n was connected to a central member of stage- through a torsional spring) representing a permanent or clutch connection.

An example set of permanent or clutch stage couplings were shown in solid lines, while the model allows any user defined set of couplings. The torsional dynamic model of the planetary gear set of stage-n was shown. That model was similar to the one proposed by Al-shyyab and Kahraman with the exception of torsional springs of stiffnesses added there to represent coupling of the central member j of stage-n with the central member of stage-m. That was discrete model employs a number of simplifying assumptions as discussed in detail in reference.

The central elements were constrained by torsional linear springs of stiffnesses respectively. A particular central member was held stationary by assigning a very large constraint stiffness value to it. Likewise, a zero value for the constraint stiffness indicates that was central member was not connected to the housing. The magnitudes of the stage coupling stiffness's chosen to represent the torsional stiffness of the actual (24)

R. G. Parker et al. (2000) studied the dynamic response of a helicopter planetary gear system was examined over a wide range of operating speeds and torques. The analysis tool was a unique, semi analytical finite element formulation that admits precise representation of the tooth geometry and contact forces that were crucial in gear dynamics. Importantly, no a priori specification of static transmission error excitation or mesh frequency variation was required; the dynamic contact forces were evaluated internally at each time step. The calculated response shows classical resonances when a harmonic of mesh frequency coincides with a natural frequency. However, particular behavior occurs where resonances expected to be excited at a given speed were absent. That absence of particular modes was explained by analytical relationships that depend on the planetary configuration and mesh frequency harmonic.

The torque sensitivity of the dynamic response was examined and compared to static analyses. Rotational mode response was shown to be more sensitive to input torque than translational mode response.

Planetary gears yield several advantages over conventional parallel shaft gear systems. They produce high speed reductions in compact spaces, greater load sharing, higher torque to weight ratio, diminished bearing loads and reduced noise and vibration. They were used in automobiles, helicopters, aircraft engines, heavy machinery, and a variety of other applications. Despite their advantages, the noise induced by the vibration of planetary gear systems remains a key concern. In helicopters, for example, cabin Noise exceeding 100 dB was directly traceable to the last stage planetary gear mounted to the cabin.

Planetary gears have received considerably less research attention than single mesh gear pairs. There was a particular scarcity of dynamic response calculations. The purpose of that work was to characterize the dynamic response of a planetary gear system under a wide range of operating conditions.

The analytical technique combines a unique, semi-analytical finite element approach with detailed contact modeling at the tooth mesh. That approach was specifically developed to examine the mechanics of precisely- machined, contacting elastic bodies such as gears. The semi- analytical finite element formulation does not require a highly refined mesh at the contacting tooth surfaces. That dramatically reduces the computational effort and allows calculation of the Dynamic response at a sufficient number of time steps to study the response in the frequency domain. In contrast, the need

For extremely refined gear tooth meshes limits conventional finite element analysis to static analyses and free vibration Eigen solutions. The current analysis provides a more accurate and comprehensive study of planetary gear dynamic response than was reasonably possible, or has been conducted, with conventional finite element analysis. (8)

Summary

Many of the researchers did the work on vibration and noise in gear trains. Because the vibration and noise are the major aspect in automobile as well as industrial sector. To reduce the vibration and noise they used many methods like changing the number of teeth, by using ansys software, analytically, stiffness of the gear.

Research Gap

The researchers did the study vibration and noise in gear pair and reduced it by various methods. The researchers vary the tooth profile for better performance. The majority of the researchers were focused on stiffness of gear, changing the tooth shape, and geometrical parameter, phasing of gear. While system is in continuous operation this methods fails to reduce the vibration in gear train. There was a very few research did on the planetary gear train to reduce the vibration.

CHAPTER 3

MATERIAL SELECTION FOR GEAR & THEIR PROPERTIES

3.1 Selection of Plastic Materials

When selecting plastic gear materials, look for those with sufficient strength and stiffness to handle the expected loads. Then make sure that any changes in dimensions and frictional characteristics due to environmental conditions are acceptable for the application. Material suppliers can provide a lot of useful information on these properties for various plastics.

But you may still need to test prototype gears under realistic operating and environmental conditions to verify that they'll perform as intended. Gear teeth must have adequate strength and fatigue endurance to carry normal and shock loads. Fatigue data from gear tests will help verify adequate bending fatigue strength at the tooth root. Fatigue or S-N curves typically show root bending stress as a function of the number of operating cycles. These curves should show the effects of temperature.

The stiffness and deflection of meshing teeth depend on the elastic modulus. A high modulus (rigidity) minimizes tooth deflection, whereas a low modulus reduces shock loads and noise. The elastic modulus depends largely on temperature, loading rate, humidity, and chemical exposure. Tensile tests or Dynamic Mechanical Analysis (DMA) measurements give temperature-related data. But the effects of the other variables are not as well defined, and usually require testing.

Dimensionally stable materials help maintain gear tooth contact ratio, tooth tip clearance, and overall geometry. Factors that affect plastic part dimensions include temperature, moisture, chemical exposure, and manufacturing process.

The coefficient of linear thermal expansion is commonly available, and ranges widely for different materials. But it may not accurately represent temperature-induced dimensional changes. Therefore, you may need to conduct tests to determine these changes. Moisture data are usually expressed as the percent of moisture absorption rather than as a dimensional change. Dividing the percent of moisture absorbed by four gives the approximate dimensional change in percent. Because plastic gears may encounter a

Variety of chemicals and lubricants, the effects of specific chemicals often require testing. Wear is a primary concern for plastic gears that run without lubrication. These gears tend to run hot at low loads, causing tooth wear failures. Therefore, they should be made from materials with high wear resistance and good lubricity.. Wear rate may be affected by temperature and moisture, so gears must be tested under these application conditions.

3.2 Why Plastic?

For drive designers, plastic gears offer several benefits including design flexibility, reduced noise, and the ability to operate without lubrication. Other benefits include lower cost and weight, higher efficiency, (improved accuracy), and chemical resistance.

Design flexibility. The design opportunities that plastic gears afford are a major advantage. They can be molded in shapes difficult to machine in metal..

Cost. They are molded in high volumes at low cost, typically one-half to one-tenth that of Stamped, machined, or powder metal gears. They're usually ready to use as-molded and require no finishing.

Weight. Plastics inherently weigh less, typically 15 to 20% as much as steel, but plastic Gears must be larger to transmit the same power.

Noise. Compared with metal, plastic gears run much quieter. Depending on the application, their sound level compares to that of metal gears. Plastic gear teeth deform, compensating for noise-producing gear misalignment and tooth errors, and the material absorbs impacts that would otherwise cause running noise.

Efficiency. A low coefficient of friction means less horsepower wasted in heat. Plastic gears also lend themselves to efficient designs such as split-path planetary drives.

Lubrication. Inherent lubricity and chemical resistance mean plastic gears can be used with or without lubrication as needed for specific applications.

Accuracy. With consistent material quality, and accurate molding process control, plastic gears can achieve high precision.

Durability. Chemical and corrosion resistance typically exceeds that of metal gears, Especially in wet applications.

3.3 Mechanical Properties of Various Plastics

Material	UTS (MPa)	E (GPa)	Elongation (%)	Poisson's ratio (ν)
ABS	28–55	1.4–2.8	75–5	—
ABS, reinforced	100	7.5	—	0.35
Acetal	55–70	1.4–3.5	75–25	—
Acetal, reinforced	135	10	—	0.35–0.40
Acrylic	40–75	1.4–3.5	50–5	—
Cellulosic	10–48	0.4–1.4	100–5	—
Epoxy	35–140	3.5–17	10–1	—
Epoxy, reinforced	70–1400	21–52	4–2	—
Fluorocarbon	7–48	0.7–2	300–100	0.46–0.48
Nylon	55–83	1.4–2.8	200–60	0.32–0.40
Nylon, reinforced	70–210	2–10	10–1	—
Phenolic	28–70	2.8–21	2–0	—
Polycarbonate	55–70	2.5–3	125–10	0.38
Polycarbonate, reinforced	110	6	6–4	—
Polyester	55	2	300–5	0.38
Polyester, reinforced	110–160	8.3–12	3–1	—
Polyethylene	7–40	0.1–1.4	1000–15	0.46
Polypropylene	20–35	0.7–1.2	500–10	—
Polypropylene, reinforced	40–100	3.5–6	4–2	—
Polystyrene	14–83	1.4–4	60–1	0.35
Polyvinyl chloride	7–55	0.014–4	450–40	—

3.4 Properties of Nylon-6

1. Good mechanical and abrasion resistance
2. Lubricating resistant to most chemicals but it absorbs water.

3.5 Applications of Nylon-6

1. **Mechanical Components**: - Gears, bearings, rollers, bushings, fasteners, guides, zippers, surgical equipment's.

Summary

This chapter is related with the properties of various plastic materials. Nylon-6 is selected for manufacturing of planetary gear as this material is having good mechanical properties and resistance to chemicals.

CHAPTER 4

EXPERIMENTAL SET-UP AND MEASUREMENTS

As per the requirement to accomplish the objectives it was necessary to develop the method to reduce PGT noise and vibrations by gear itself without requiring the additional energy, actuators, and advanced signal processing techniques. Viewing this need the method of noise reduction in planetary gears by phasing is introduced in this research work.

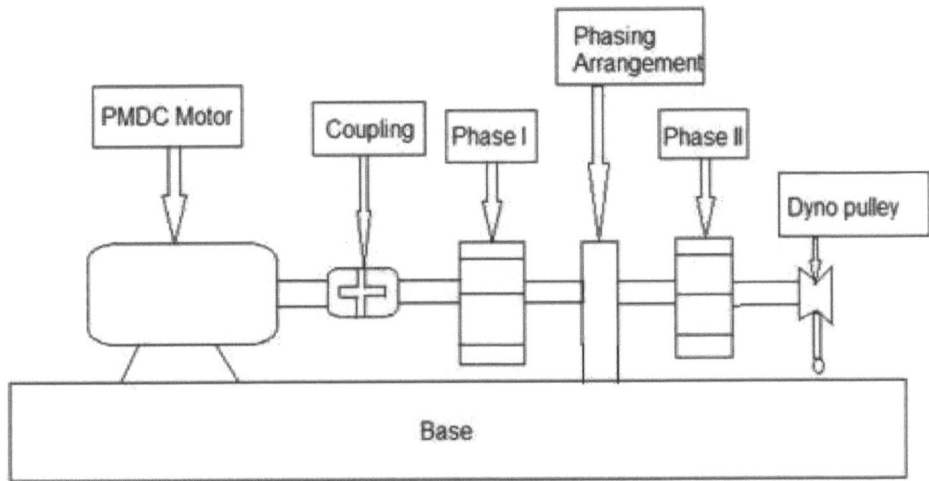

Figure 5: Schematic layout of test set-up for measurement of planetary gear set noise and vibration

In order to study the effect of phasing on noise and vibrations of planetary gear set the required experimental set-up was developed as shown in Figure 5.

Figure 5 shows schematic layout of test set-up developed for the measurement of noise level of planetary gear set by phasing. Figure 5 also shows the position of various components like motor, planetary gear sets 1 and 2, coupling, and speed regulator. The experimental work was carried out to study the effect of meshing phasing on noise level and vibrations of Nylon-6 planetary gear set. For this purpose experimental set-up was built

As shown in Figure 5. Rectangular plate is placed between planetary gear sets 1 and 2 to provide meshing phase difference between ring gear of gear sets 1 and 2. Noise level is measured for two different arrangements as with phasing and without phasing. Experimental set-up shown in Figure 5 consists of different components such as PMDC motor, love joy coupling, planetary gear set, and speed selector which are explained in this section.

4.1 Motor Selection

The speed reduction is to be achieved using two stage reductions; hence the stage wise reduction of the system is as follows. Figure 06 shows the PMDC motor as shown in figure Stage-I

(a)   Input speed = 1440 rpm.

(b)   Reduction ratio = 4.

(c)   Output speed of stage I = 1440/4 = 360 rpm.

Figure 06: Permanent magnet DC motor

Stage-II

(d) Input speed = 360 rpm.

(e) Reduction ratio = 4.

(f) Output speed of stage-II = 360/4 = 90 rpm.

(g) Hence, maximum motor speed = 1500 rpm.

Power = 2 ×π× 1500 × 0.5/60 = 78.5W.

Hence motor of 90 watt is selected as shown in Figure4.

(i) Motor type: fractional HP permanent magnet DC motor.

(ii) Torque: 5 kgcm.

(iii) Speed: 1500 rpm.

(iv) Input power: 90 watt.

4.2 Lovejoy Coupling

Lovejoy Coupling L-075 (as per specification shown in Table 1) is selected for the given Application with outside diameter of hub (Do) = 38 mm Inside diameter of hub (Di) = 12mm

Lovejoy Coupling L-075 considered being a hollow shaft subjected to torsional load.

Figure 07: Love-joy Coupling

Figure 7 shows the Love Joy coupling which is used in the setup for couple the two shaft that is of motor shaft and another shaft on which the gear is mounted. The diameter of the coupling are given above.

Table 1: Material selection for coupling

Designation	Ultimate tensile strength N/mm2	Yield strength N/mm2
EN 6	600	480

4.3 Selection of Gear Box

Nylon-6 planetary gear box (Table 2) is selected based on mechanical properties, wear resistance, lubrication and material availability, torque, and other parameters.

Gear Pair-1 (Phase I).

Tables 3 and 4 show the planet gear and internal gear ring specifications.

Gear Pair-2 (Phase I).

Tables 3 and 5 show the planet gear and sun gear specification.

4.4 Coupler Shaft

Coupler shaft is used to connect planetary gear sets one and two. Table 6 shows different mechanical properties of selected material EN24.

4.5 Selection of Gear Box
Gear Pair-1& II (Phase II)

Tables 3 and 4 show the planet gear and internal gear ring specifications.

4.6 Gear Pair-3 (Phase II)

Tables 8 and 9 show the specification of planet gear and internal gear ring.

4.7 Gear Pair-4 (Phase II)

Tables 7 and 8 show the planet gear and internal gear ring specification.

4.8 Sound Level Meter

For measurement of noise level in PGT sound level meter with specifications as shown in Table 12.

4.9 FFT Analyzer

For measurement of noise level in PGT FFT Analyzer with specifications as shown in Table 13 was used.

Table 2: Mechanical properties of Nylon-6 gear box

Mechanical properties ($73°F$)	ASTM test method	Units	Nylon 6/6	Nylon 6/6 GF30
Tensile strength	D638	Psi	12,400	27,000
Elongation	D638	%	90	3
Flexural strength,	D790	Psi	17,000	39,100
Flexural modulus,	D790	Psi	4.1×10^5	12×10^5
Izod impact strength, Notched,	D256	-	R120-M79	M101
Rockwell hardness	D785	ft-lbs/in.	1.2	2.1

Table 3: Planet gear specification

Material	Nylon-6
Module	1.375 mm
Number of teeth	16
Addendum diameter	24.75 mm
Pitch circle diameter	22 mm

Table 4: Internal gear ring specification

Material	Nylon-6
Module	1.375 mm
Number of teeth	48
Addendum diameter	68.75 mm
Pitch circle diameter	66 mm

Figure 8: Internal ring gear of nylon-6

Figure 8 shows the internal ring gear made of nylon 6 materials and shows the number of teeth of ring gear. The ring gear is made up of nylon material which is shown in figure 8.

Table 5: Sun gear specification

Module	1.375 mm
Number of teeth	16
Addendum diameter	24.75 mm
Pitch circle diameter	22 mm

Table 6: Selection of material for coupler shaft

Designation	Ultimate tensile strength N/mm2	Yield strength N/mm2
EN24 (40 N; 2 Cr 1 Mo 28)	720	600

4.10 Phasing Arrangement

Planetary gear set used in experimental set-up. This planetary gear set is used to calculate the angle of indexing. Phase difference is provided in between ring gear of planetary gear sets 1 and 2. Figure 9 show without phasing and with phasing arrangement. Phase difference is provided as shown in Figure 9.

No of teeth of ring gear = 48.

Angle of pitch = 360/48 = 7.5^0 = 7.5/2 = 3.75^0.

Angle of indexing = 5 × angle between two teeth = 5 × 3.75^0 = 18.75^0.

Figure 9: Phasing arrangement

Table 7: Planet gear specification

Material	Nylon-6
Module	1.375 mm
Number of teeth	16
Addendum diameter	24.75 mm
Pitch circle diameter	22 mm

Table 8: Internal gear specification

Material	Nylon-6
Module	1.375 mm
Number of teeth	48
Addendum diameter	68.75 mm
Pitch circle diameter	66 mm

Table 9: Planet gear specification

Material	Nylon-6
Module	1.375 mm
Number of teeth	16
Addendum diameter	24.75 mm
Pitch circle diameter	22 mm

Table 10: Sun gear specification

Material	Nylon-6
Module	1.375 mm
Number of teeth	16
Addendum diameter	24.75 mm
Pitch circle diameter	22 mm

Table 11: Specifications of sound level meter

Frequency range	31.5Hz ~8KHz
Display	LCD
Measuring level range	35 ~ 130 dB
Accuracy	±1.5 Db (under reference conditions)
Dynamic range	65 dB
Power supply	One 9V battery, 006P or IEC 6F22 or NEDA 1604
Calibration	Electrical calibration with the internal oscillator (1 kHz sine wave)

Table 12: Specifications of FFT Analyzer

Model	COCO-80 (gram M/s Crystal Instrument, USA)
Accelerometer	3 numbers ABRO Inc USA make accelerometer sensor type
Frequency range	Up to 84 kHz analysis frequency (192 k samples per second)
Voltage ranges	±0.01, ±0.1, ±1.0, ±10V
Resolution	24-bit
Dynamic range	115 dBfs two-tone test, 100 linear averages
Accuracy	±0.04 dB (1 kHz sine at full scale)

Figure 10: FFT analyzer

Figure 10 shows the FFT analyzer which is used for experimentation while taking readings of vibration and finding out the spectrum obtained from the FFT analyzer.

4.11 Noise Measurement in PGT

Noise measurement and signal analysis are important tools when experimentally investigating gear noise. Gears create noise at specific frequencies, related to the rotational speed and number of teeth of the gear. It is also possible to detect different errors like for example, run out (eccentricity) due to side-band generation [22]. Closely related is also vibration measurement and signal analysis for the purpose of gear fault detection, used in machine diagnostics in order to detect gear failures before catastrophic failure occurs. Middleton [23] discussed noise testing of gearboxes in the production line in which a noise testing equipment, utilizing low cost digital analysis and control techniques, was described. For each gear of the gearbox, the speed was ramped up while measuring noise with three microphones. For each order of interest (gear mesh frequency and its harmonics) the pass/fail target levels were defined by testing a selection of gearboxes which had noise characteristics regarded as just acceptable.

A test rig was developed by Gielisch and Heitmann [24] to investigate gear noise from a car rear axle, without the need for a complete vehicle. Vibrations were measured on the final drive casing and the corresponding forces and torques in the gearing were calculated. The investigation gave information about the dynamics of the driving gear and the possibility to make comparisons between different driving gears. Oswald et al. [25] investigated the influence of gear design on gearbox radiated noise in which nine different spur and helical gear designs were tested in a gear noise test rig to compare the noise radiated from the gearbox top for the various gear design and the results were summarized as follows.

1. The total contact ratio was the most significant factor for reducing noise, increasing either the profile or face contact ratio reduced the noise.
2. The non-involutes spur gears were 3-4 dB noisier than in volute spur gears.
3. High contact ratio spur gears showed a noise reduction of about 2 dB over standard spur gears.
4. The noise level of double helical gears averaged about 4 dB higher than otherwise similar single helical gears.

In this paper method proposed in [7] is extended by applying novelty of phasing concept to reduce the noise level and resulting vibrations of planetary gear set without the requirement of additional instruments like actuators, external power, and signal processing techniques which generally results in increase in cost.

Summary

This chapter is related with the experimental setup by selecting the various components. This chapter gives the specification of each component's and method of phasing.

CHAPTER 5
RESULTS AND DISCUSSION

5.1 Noise Measurement in single PGT, Without Phasing & With Phasing arrangement

After finalizing the model we have to take the readings of noise on noise measuring instrument. The Noise is calculated on Sound level meter and obtained the results are as follows.

Table 13: Noise Measurement in single PGT arrangement

Speed (rpm)	Noise (dB)
100	36
200	38
300	40
400	42
500	45
600	48
700	49
800	51
900	54

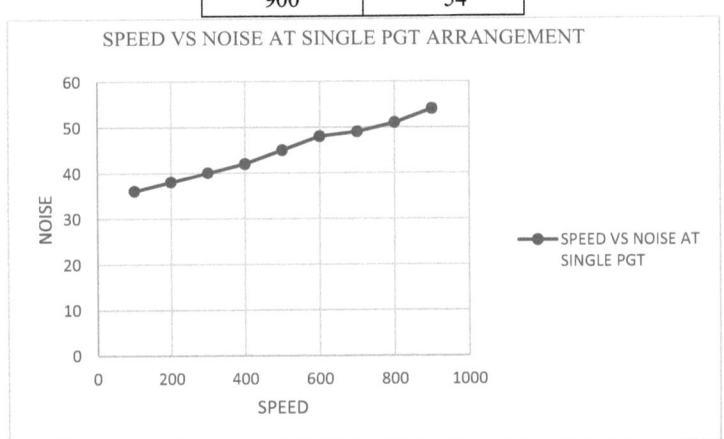

Graph 1: Graph of Speed Vs Noise at single PGT arrangement

This graph 1 is related with the noise measurement in PGT at single PGT arrangement. The speed of the motor is varies between 100 to 900 RPM.

Table 13: Noise Measurement at without phasing arrangement

Speed (rpm)	Noise (dB)
100	38
200	40
300	43
400	46
500	49
600	49
700	51
800	53
900	55

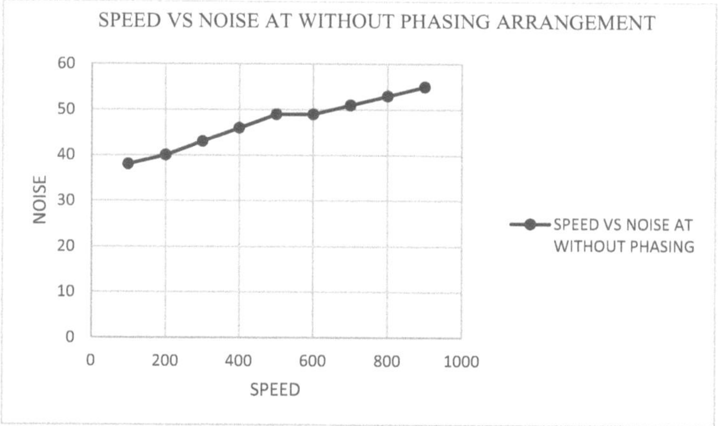

Graph 2: Graph of Speed Vs Noise at without phasing arrangement

This graph is related with the noise measurement in PGT at without phasing arrangement. The speed of the motor is varies between 100 to 900 RPM.

Table 14: Noise Measurement with Phasing Arrangement

Speed	Noise (dB)
100	32
200	34
300	36
400	39
500	42
600	44
700	46
800	47
900	49

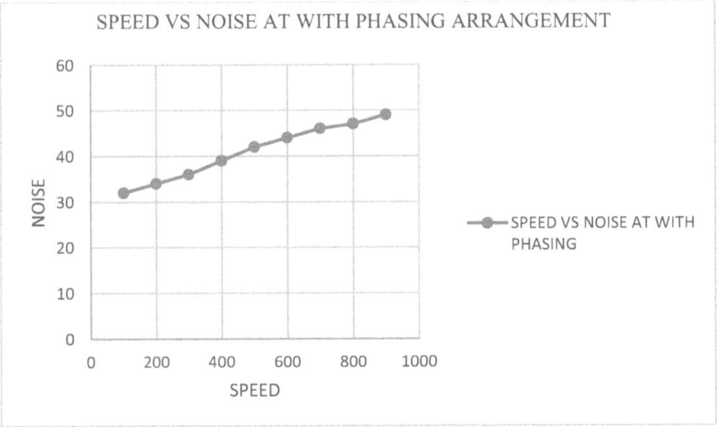

Graph 3: Graph of Speed Vs Noise with phasing

This graph is related with the noise measurement in PGT at with phasing arrangement. The speed of the motor is varies between 100 to 900 RPM.

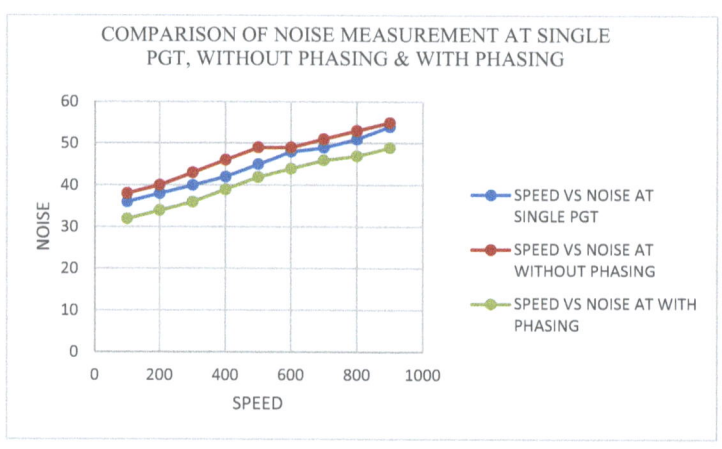

Graph 4: Graph of Speed Vs Noise at single PGT, with phasing & without phasing arrangement

This Graph gives the graph for Speed Vs Noise at single PGT, phasing & without phasing arrangement. The graph is plotted at 100 to 900 RPM and noise is varying at different speeds. This proves that the noise is reduces at phasing arrangement between the gear pair.

Summary

The noise in the planetary gear train reduces by phasing arrangement as compared to without phasing arrangement and single PGT arrangement. This arrangement reduces the noise by 7 % to 9 %.

5.2 Vibration measurement in single planetary gear trains

The vibration in the single planetary gear train is measured with the help of FFT analyzer. The vibration is measured in terms of displacement readings which are obtained from FFT analyzer.

Table 15: Acceleration, Velocity & Displacement values for single PGT

Speed	Acceleration (m/s^2)	Velocity (mm/s)	Displacement (μm)
100	1.66	6.6	31
200	1.88	5.7	40
300	2.11	8.3	75
400	2.62	15.9	155
500	3.13	21.5	180

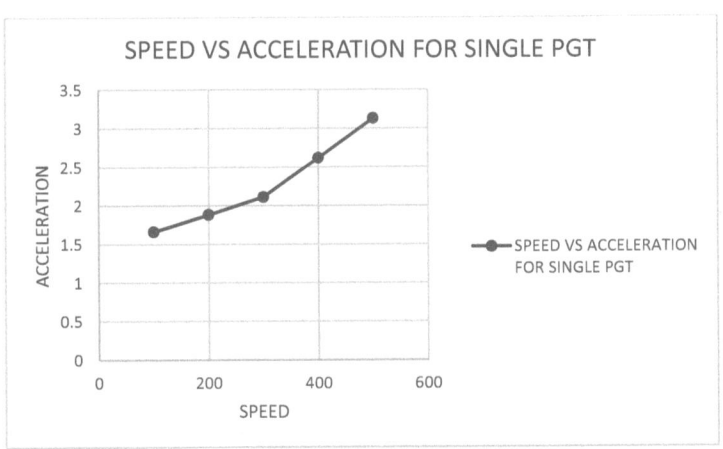

Graph 5: Graph of Speed VS Acceleration for single PGT

This graph is plotted between the Speed VS Acceleration for the Single PGT. This graph gives the accelerations for the various speed of motor.

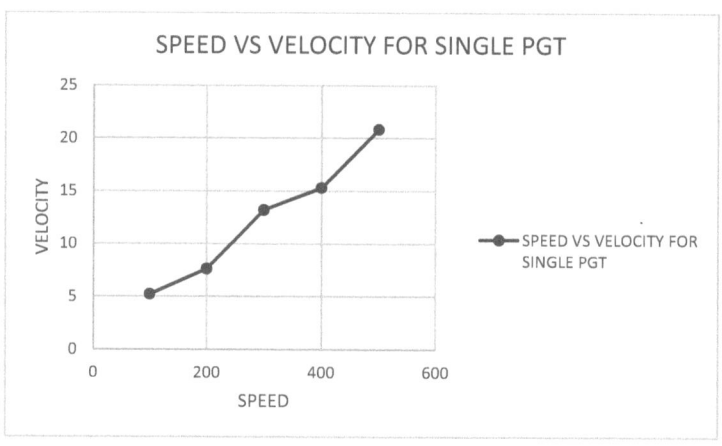

Graph 6: Graph of Speed VS Velocity for single PGT

This graph is plotted between the Speed VS Velocity for the Single PGT. This graph gives the Velocity for the various speed of motor.

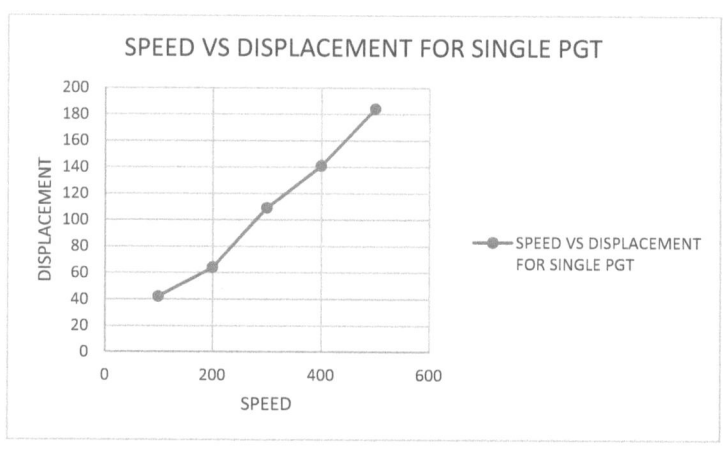

Graph 7: Graph of Speed VS Displacement for single PGT

This graph is plotted between the Speed VS Displacement for the Single PGT. This graph gives the Displacement for the various speed of motor.

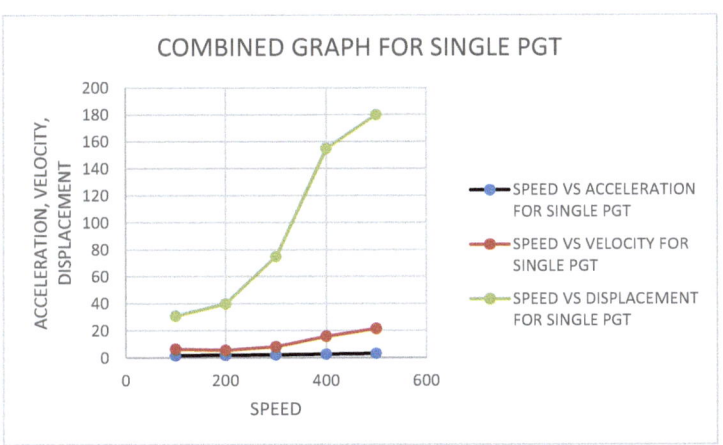

Graph 8: Combined Graph of Speed VS Displacement, Acceleration, and Velocity for single PGT

This graph is plotted between the Speed VS Displacement, Velocity & Acceleration for the Single PGT. This graph gives the values for all the velocity, displacement & acceleration for the various speed of motor.

5.3 Vibration measurement in PGT by without & with phasing of gear pair

The vibrations in the planetary gear train are measured by the without phasing arrangement between the two PGT which is measured by FFT analyzer.

Table 16: Acceleration, Velocity and Displacement Measurements At Without Phasing

Rpm	Acceleration m/s^2	Velocity mm/s	Displacement um
100	1.67	6.8	38
200	1.91	7.1	60
300	2.12	8.4	105
400	2.63	15.9	160
500	3.15	21.7	185

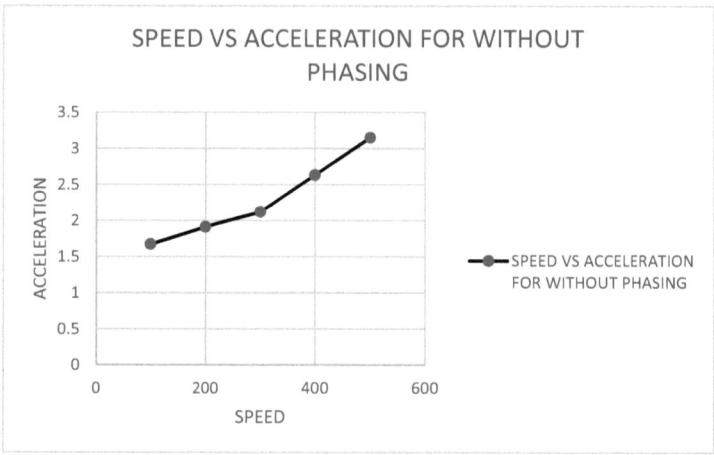

Graph 9: Graph of Speed VS Acceleration without phasing Arrangement

This graph shows the Accelerations for various speeds varying from 100 to 500 RPM for without phasing arrangement. The values of Accelerations are changes as their change in the Speed of motor.

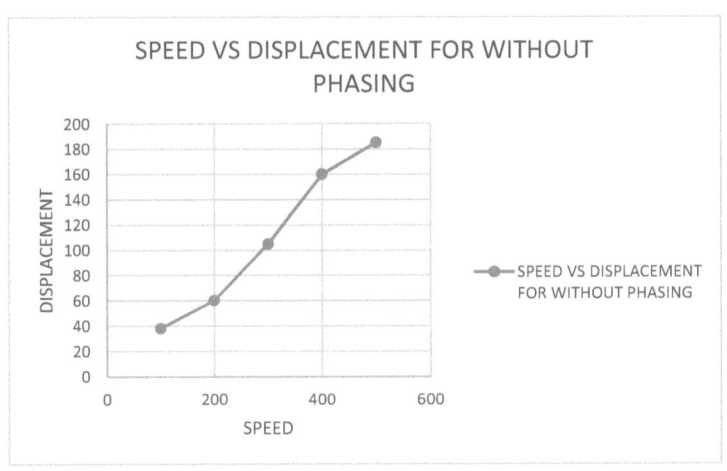

Graph 10: Graph of Speed Vs Displacement for without Phasing arrangement

This graph shows the Displacement for various speeds varying from 100 to 500 RPM for without phasing arrangement. The values of Displacement are changes as their change in the Speed of motor.

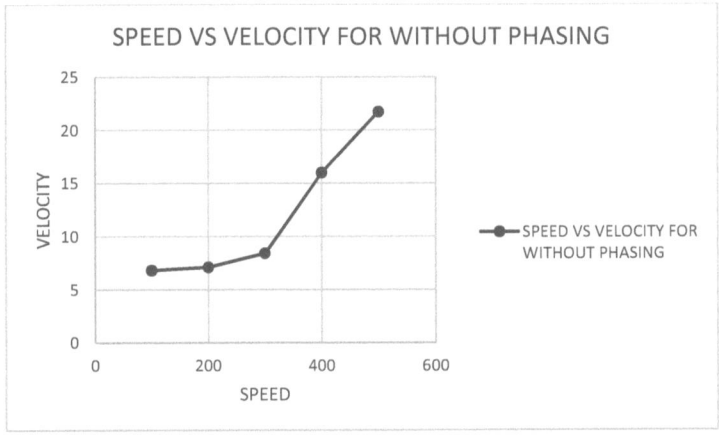

Graph 11: Graph of Speed Vs Velocity for without phasing arrangement

This graph shows the Velocity for various speeds varying from 100 to 500 RPM for without phasing arrangement. The values of Velocity are changes as their change in the Speed of motor.

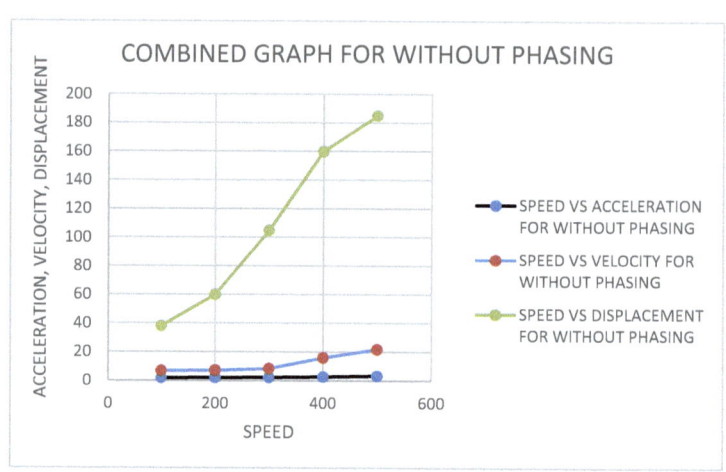

Graph 12: Combined Graph of Speed Vs Acceleration, Velocity & Displacement for without phasing arrangement

This graph shows the Acceleration, Velocity & Displacement for various speeds varying from 100 to 500 RPM for without phasing arrangement. The values of accelerations, Velocity & Displacement are changes as their change in the Speed of motor.

Table 17: Acceleration, Velocity and Displacement measurements at with Phasing Arrangement

Rpm	Acceleration m/s²	Velocity mm/s	Displacement um
100	1.16	5.8	27
200	1.86	3.1	33
300	1.18	7.2	65
400	2.61	15.8	138
500	3.002	17.3	170

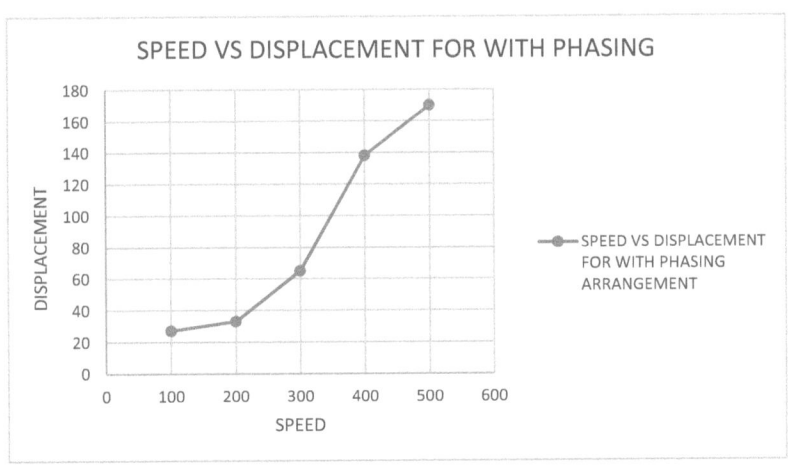

Graph 13: Graph of Speed Vs Displacement for with phasing arrangement

This graph shows the Displacement for various speeds varying from 100 to 500 RPM for with phasing arrangement. The values of Displacement are changes as their change in the Speed of motor.

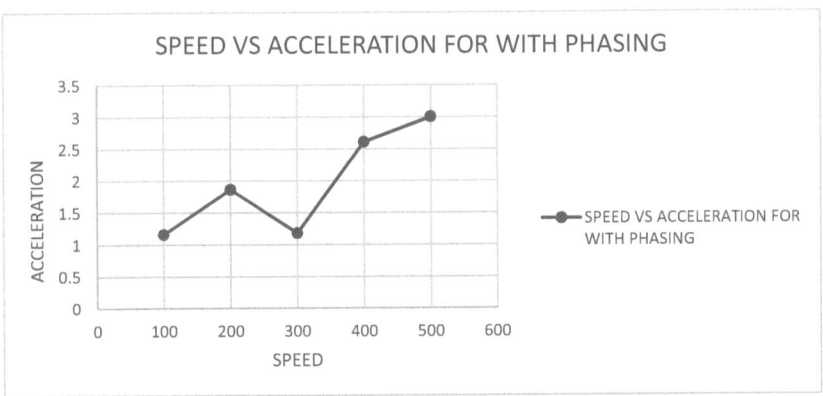

Graph 14: Graph of Speed Vs Acceleration for with phasing arrangement

This graph shows the Acceleration for various speeds varying from 100 to 500 RPM for with phasing arrangement. The values of Acceleration are changes as their change in the Speed of motor.

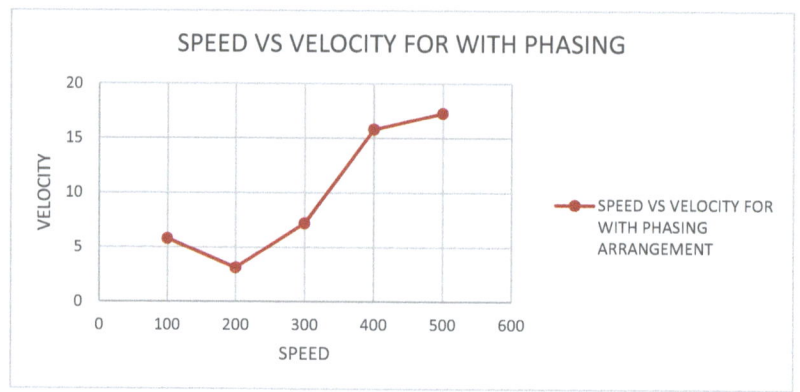

Graph 15: Graph of Speed Vs Velocity for with phasing arrangement

This graph shows the Velocity for various speeds varying from 100 to 500 RPM for with phasing arrangement. The values of Velocity are changes as their change in the Speed of motor.

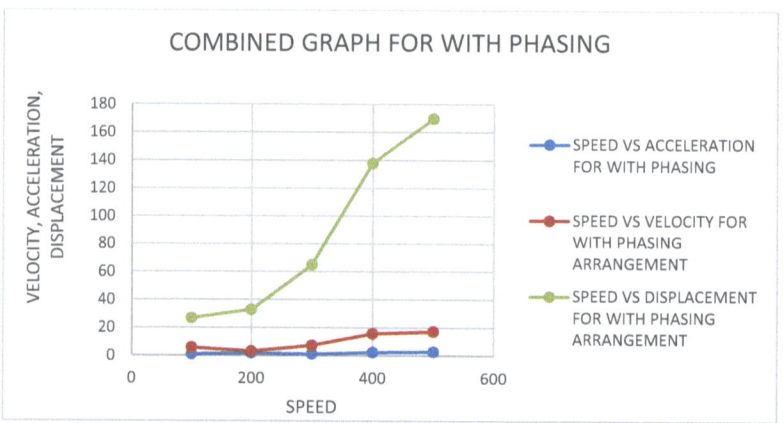

Graph 16: Combined Graph of Speed Vs Acceleration, Velocity & Displacement for with phasing arrangement

This graph shows the Acceleration, Velocity & Displacement for various speeds varying from 100 to 500 RPM for with phasing arrangement. The values of accelerations, Velocity & Displacement are changes as their change in the Speed of motor.

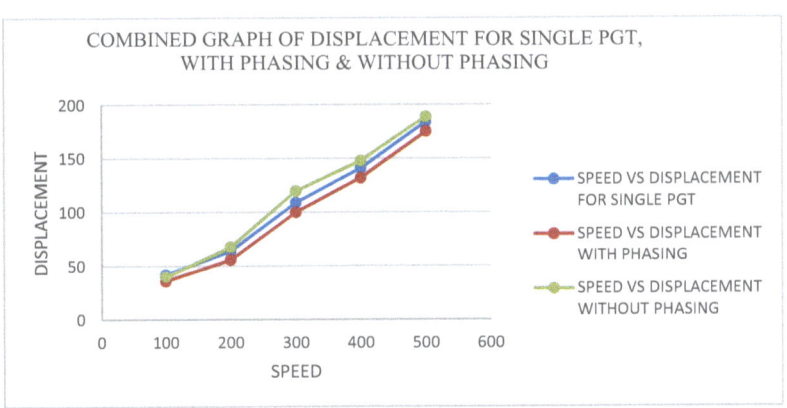

Graph 17: Comparison of Speed Vs Displacement graph for phasing, without phasing & Single PGT

This graph shows that displacement of phasing, without phasing & Single PGT arrangement. As the displacement for without phasing arrangement is more as compared to with phasing arrangement the vibration in phasing arrangement is less than that of without phasing arrangement.

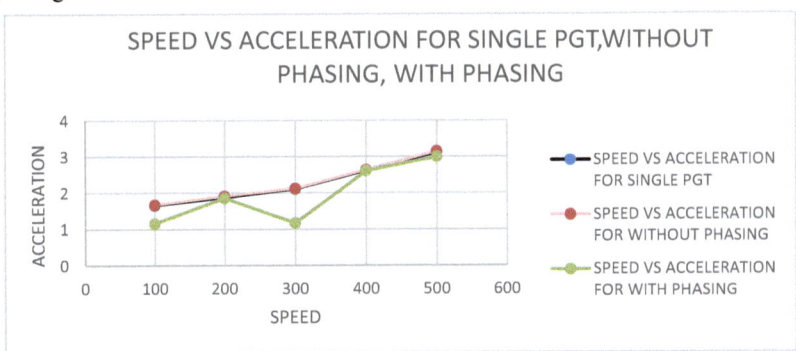

Graph 18: Comparison of Speed Vs Acceleration graph for phasing, without phasing, Single PGT arrangement

This graph shows that Acceleration of phasing, without phasing, Single PGT arrangement. As the acceleration for without phasing arrangement is more as compared to with phasing arrangement. The vibration in phasing arrangement is less than that of without phasing arrangement.

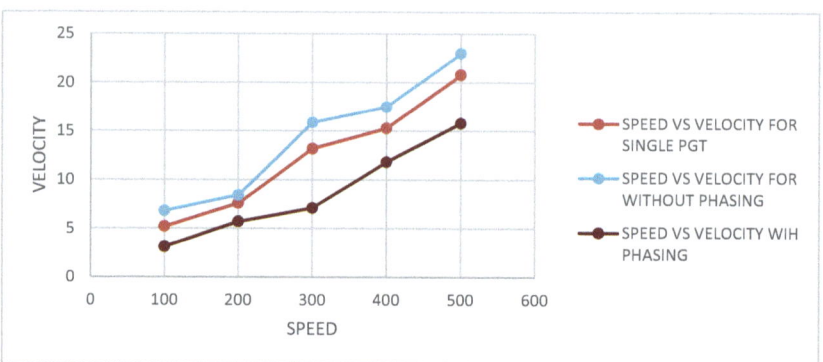

Graph 19: Comparison of Speed Vs Velocity graph for phasing, without phasing & Single PGT arrangement

This graph shows that Velocity of phasing, without phasing & Single PGT arrangement. As the Velocity for without phasing arrangement is more as compared to with phasing arrangement. The vibration in phasing arrangement is less than that of without phasing arrangement.

Summary

As the displacement of the PGT is decreases with phasing arrangement between the gear pair. This results the decrease on the vibration by phasing arrangement between the two PGT and vibration reduces up to 9% to 11%.

5.4 Acceleration Spectrums

The acceleration Spectrums obtained from the FFT analyzer for single PGT, with and without phasing arrangement at various speeds are as follows the Graph 20 to Graph 34 gives the acceleration spectrums as shown in figure.

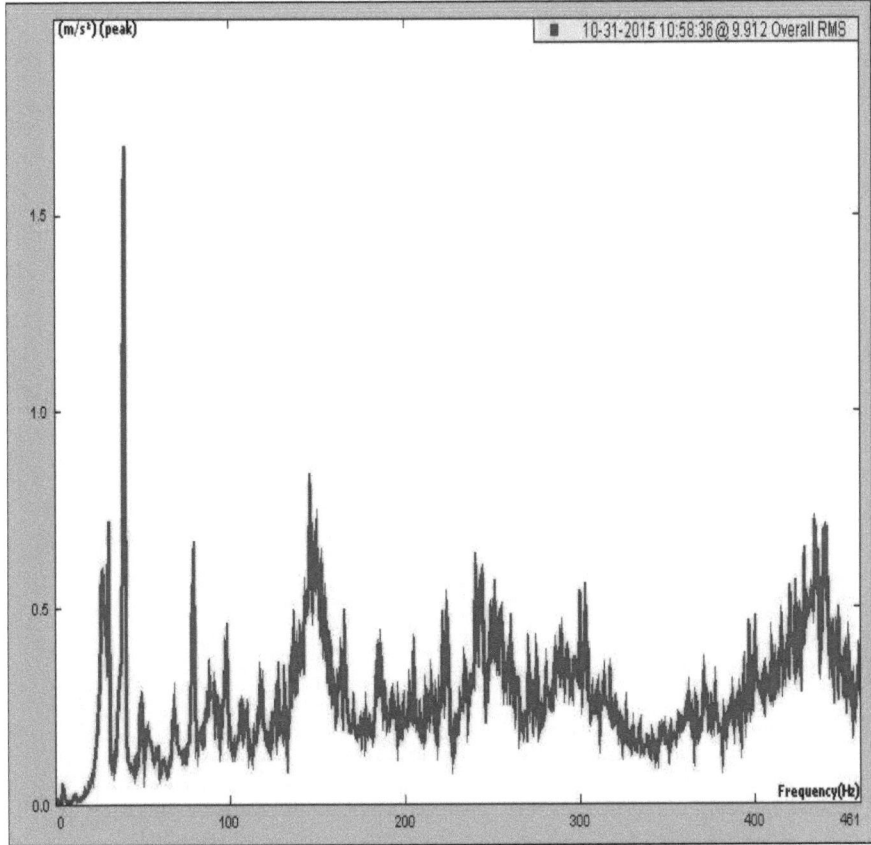

Graph 20: Acceleration Spectrum obtained from FFT analyzer

This spectrum obtained from FFT analyzer gives the acceleration spectrum at 100 RPM for single PGT arrangement. This Graph is plotted between frequency and acceleration. The natural frequency for this spectrum is 32 Hz. And that of maximum acceleration is 1.66 m/s^2.

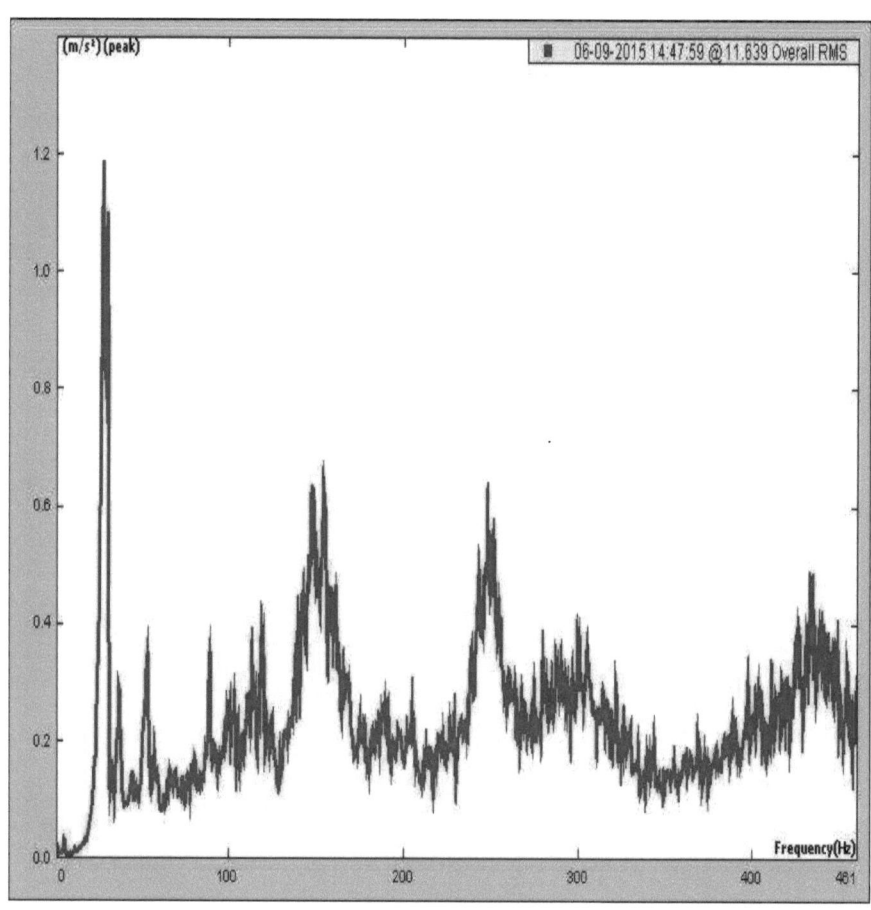

Graph 21: Acceleration Spectrum obtained from FFT analyzer

This spectrum obtained from FFT analyzer gives the acceleration spectrum at 100 RPM with Phasing arrangement. This Graph is plotted between frequency and acceleration. The natural frequency for this spectrum is 22 Hz. And that of maximum acceleration is 1.162 m/s^2.

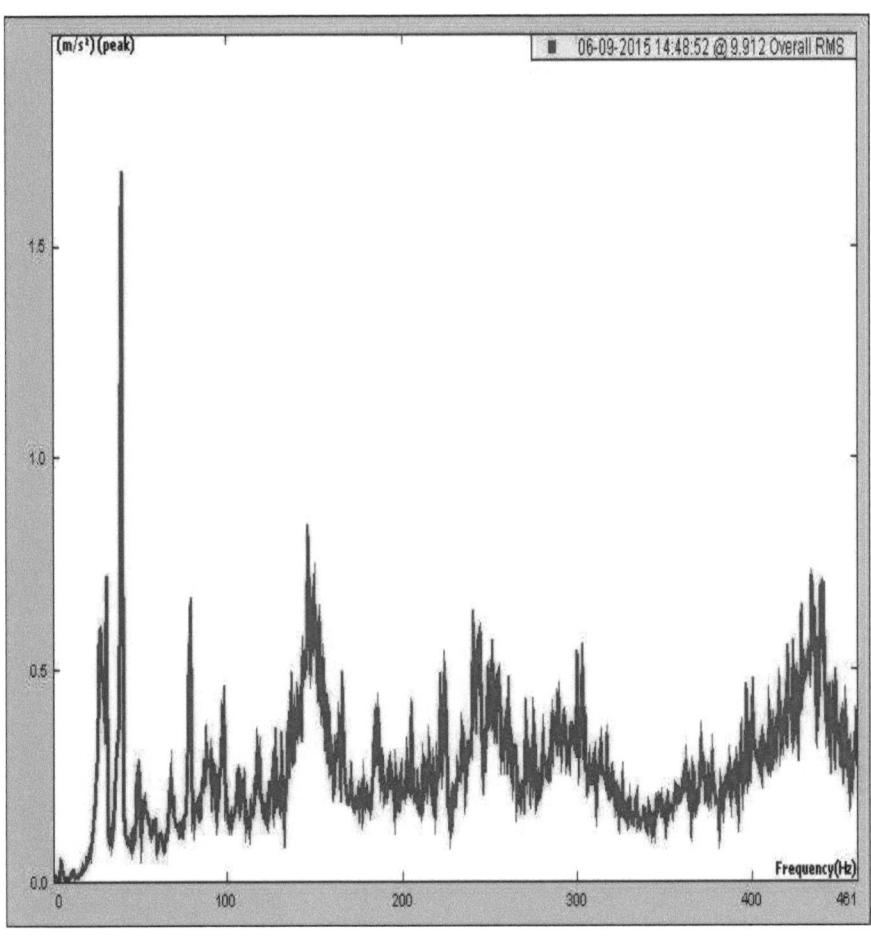

Graph 22: Acceleration Spectrum obtained from FFT analyzer

This spectrum obtained from FFT analyzer gives the acceleration spectrum at 100 RPM without Phasing arrangement. This Graph is plotted between frequency and acceleration. The natural frequency for this spectrum is 34 Hz. And that of maximum acceleration is 1.675 m/s^2.

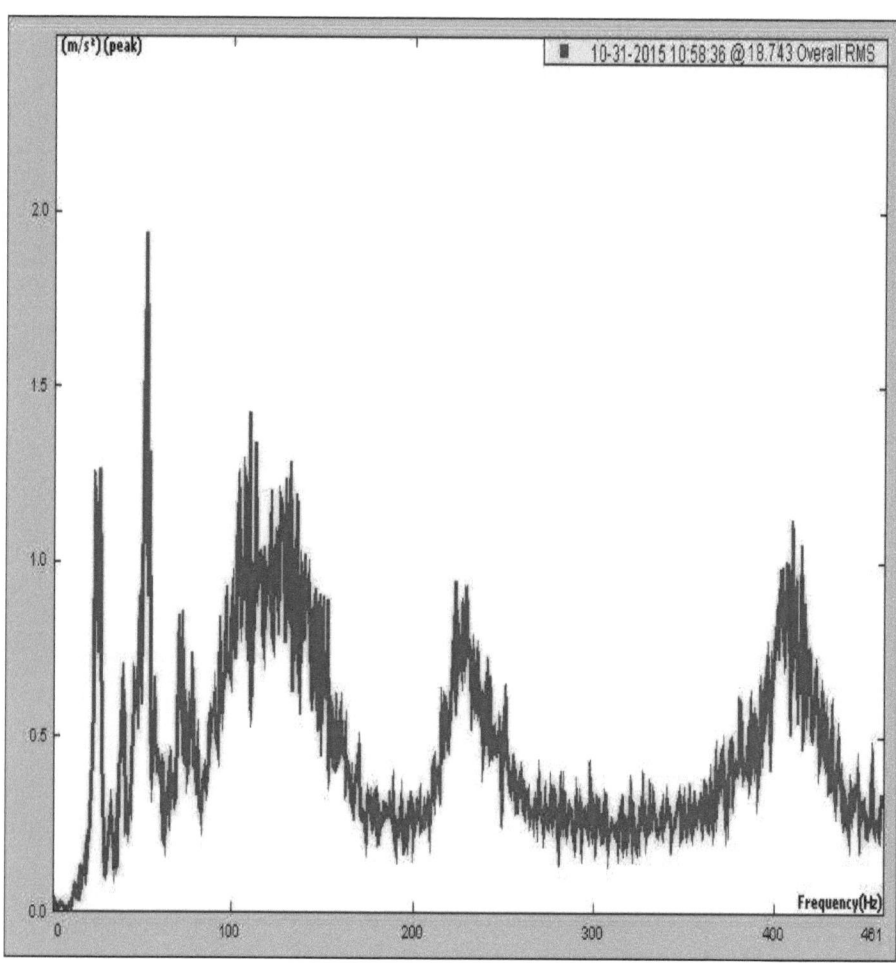

Graph 23: Acceleration Spectrum obtained from FFT analyzer

This spectrum obtained from FFT analyzer gives the acceleration spectrum at 200 RPM single PGT arrangement. This Graph is plotted between frequency and acceleration. The natural frequency for this spectrum is 48 Hz. And that of maximum acceleration is 1.882 m/s^2.

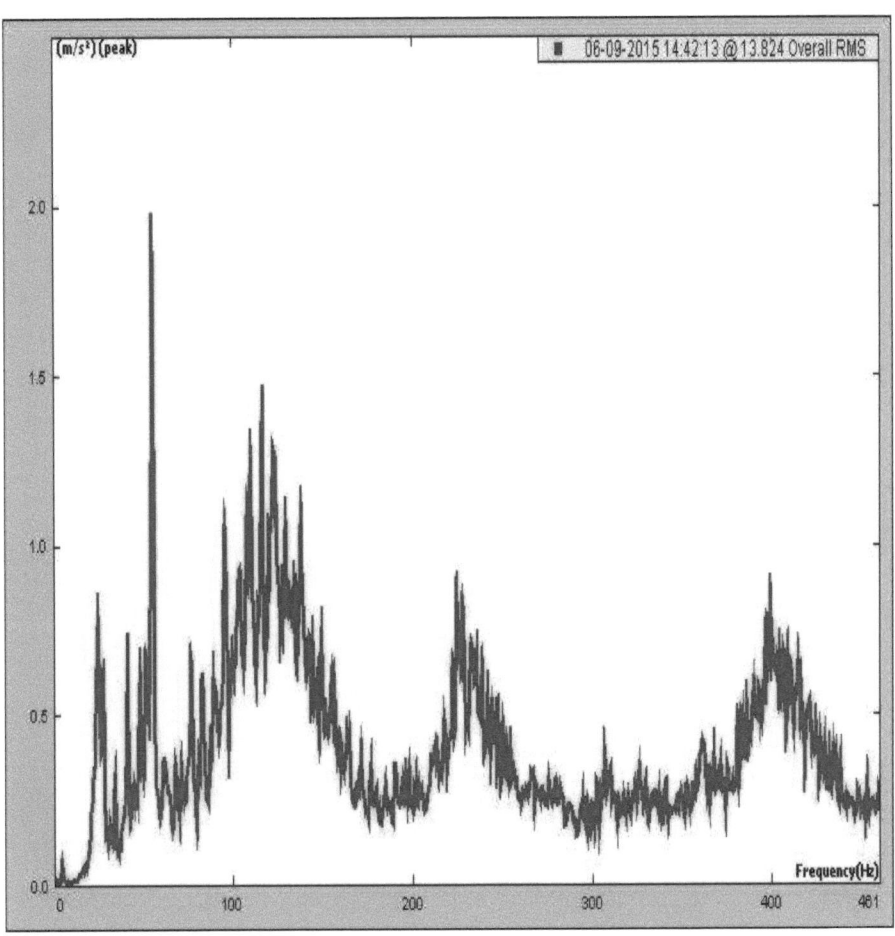

Graph 24: Acceleration Spectrum obtained from FFT analyzer

This spectrum obtained from FFT analyzer gives the acceleration spectrum at 200 RPM with Phasing arrangement. This Graph is plotted between frequency and acceleration. The natural frequency for this spectrum is 51 Hz. And that of maximum acceleration is 1.886 m/s^2.

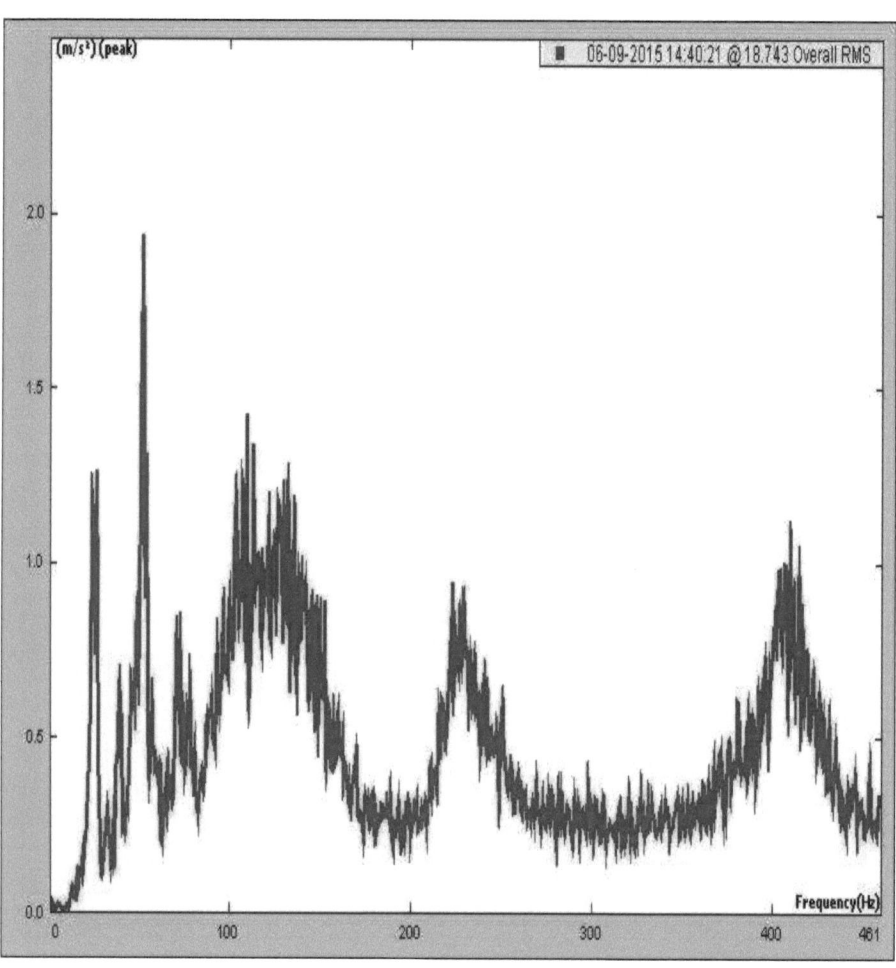

Graph 25: Acceleration Spectrum obtained from FFT analyzer

This spectrum obtained from FFT analyzer gives the acceleration spectrum at 200 RPM without Phasing arrangement. This Graph is plotted between frequency and acceleration. The natural frequency for this spectrum is 53 Hz. And that of maximum acceleration is 1.912 m/s^2.

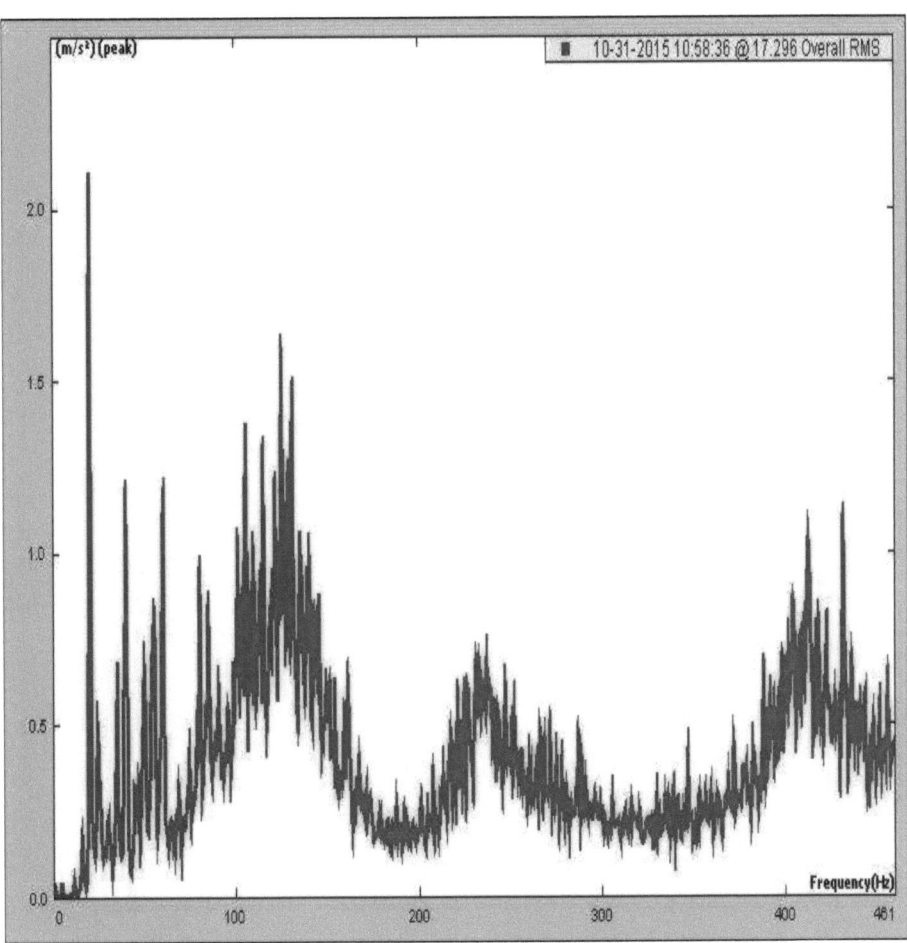

Graph 26: Acceleration Spectrum obtained from FFT analyzer

This spectrum obtained from FFT analyzer gives the acceleration spectrum at 300 RPM single PGT arrangement. This Graph is plotted between frequency and acceleration. The natural frequency for this spectrum is 22 Hz. And that of maximum acceleration is 2.114 m/s^2.

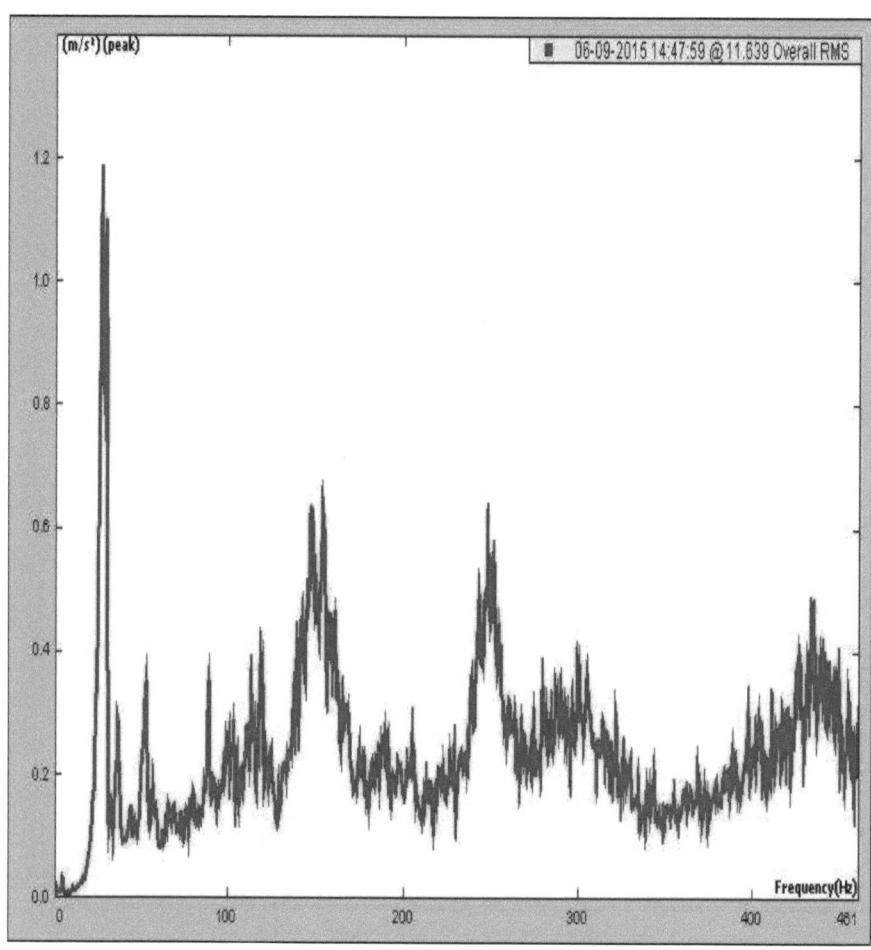

Graph 27: Acceleration Spectrum obtained from FFT analyzer

This spectrum obtained from FFT analyzer gives the acceleration spectrum at 300 RPM with Phasing arrangement. This Graph is plotted between frequency and acceleration. The natural frequency for this spectrum is 23 Hz. And that of maximum acceleration is 1.188 m/s^2.

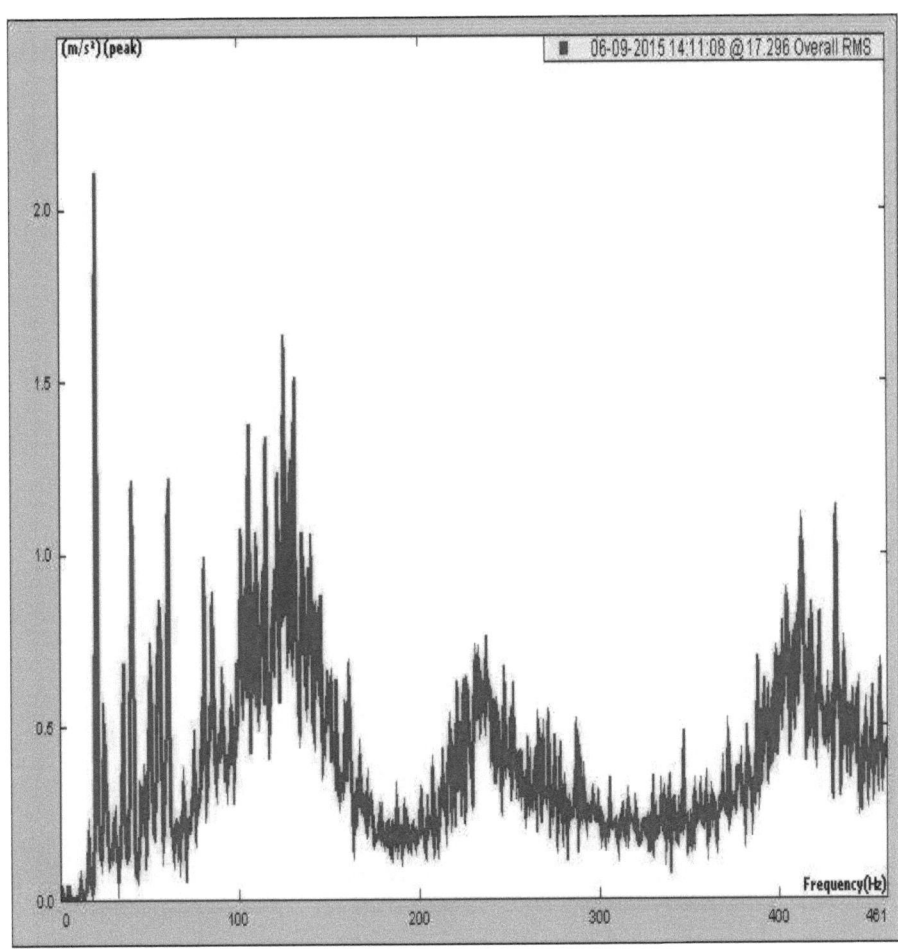

Graph 28: Acceleration Spectrum obtained from FFT analyzer

This spectrum obtained from FFT analyzer gives the acceleration spectrum at 300 RPM without Phasing arrangement. This Graph is plotted between frequency and acceleration. The natural frequency for this spectrum is 26 Hz. And that of maximum acceleration is 2.128 m/s^2.

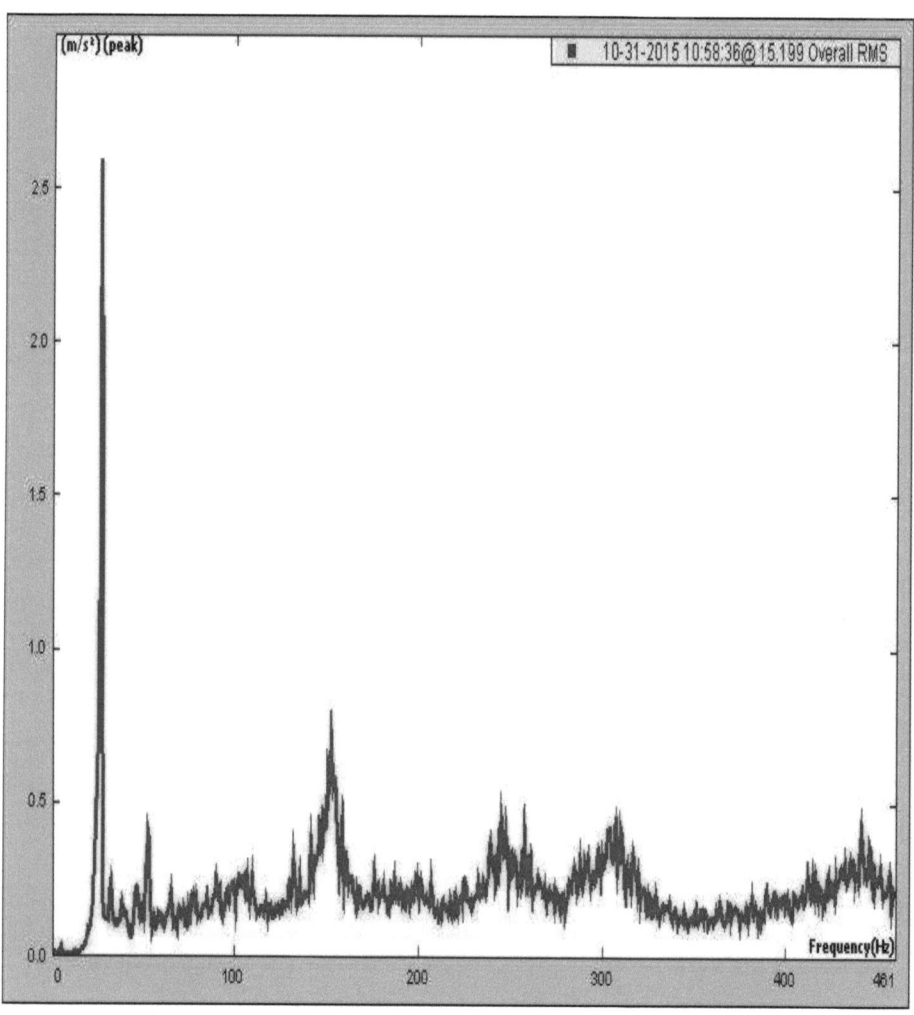

Graph 29: Acceleration Spectrum obtained from FFT analyzer

This spectrum obtained from FFT analyzer gives the acceleration spectrum at 400 RPM single PGT arrangement. This Graph is plotted between frequency and acceleration. The natural frequency for this spectrum is 26.22 Hz. And that of maximum acceleration is 2.619 m/s^2.

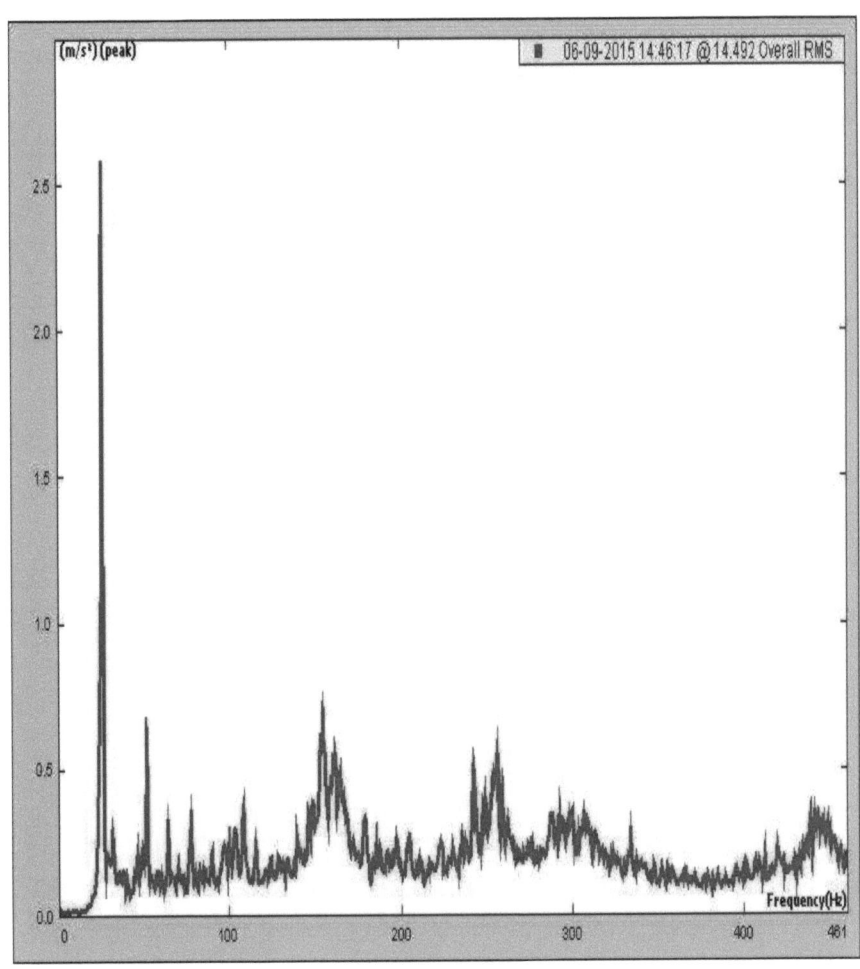

Graph 30: Acceleration Spectrum obtained from FFT analyzer

This spectrum obtained from FFT analyzer gives the acceleration spectrum at 400 RPM with Phasing arrangement. This Graph is plotted between frequency and acceleration. The natural frequency for this spectrum is 20 Hz. And that of maximum acceleration is 2.612 m/s^2.

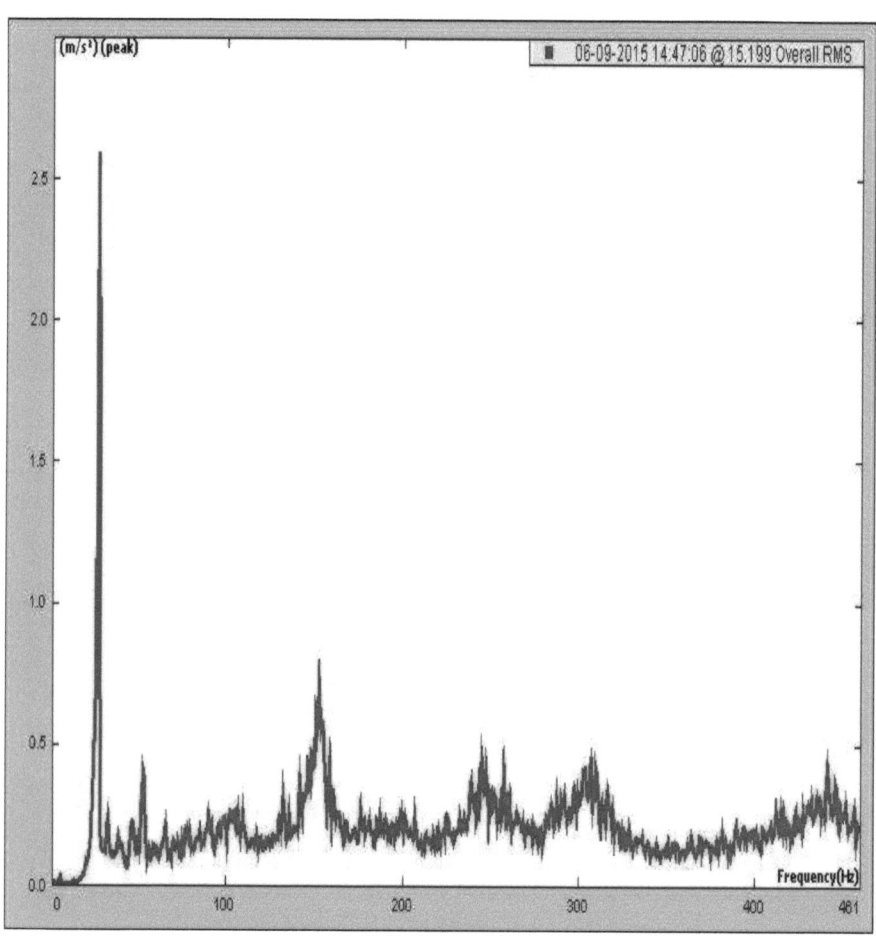

Graph 31: Acceleration Spectrum obtained from FFT analyzer

This spectrum obtained from FFT analyzer gives the acceleration spectrum at 400 RPM without Phasing arrangement. This Graph is plotted between frequency and acceleration. The natural frequency for this spectrum is 22 Hz. And that of maximum acceleration is 2.639 m/s^2.

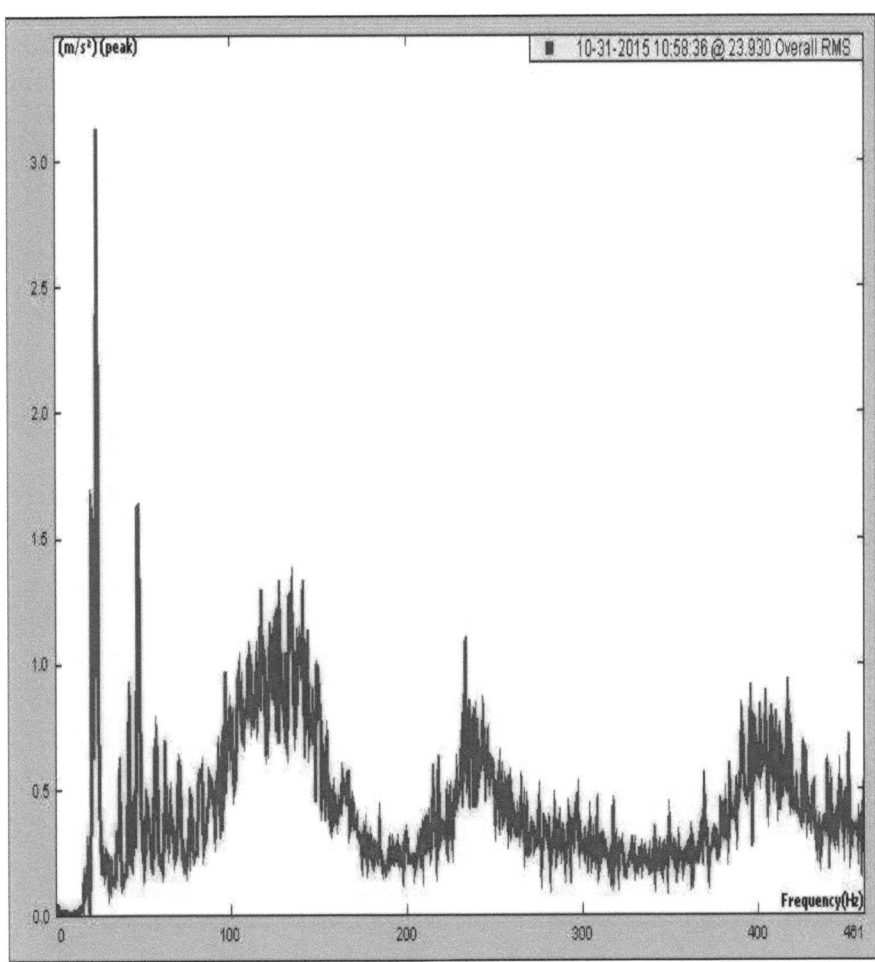

Graph 32: Acceleration Spectrum obtained from FFT analyzer

This spectrum obtained from FFT analyzer gives the acceleration spectrum at 500 RPM single PGT arrangement. This Graph is plotted between frequency and acceleration. The natural frequency for this spectrum is 20 Hz. And that of maximum acceleration is 3.130 m/s^2.

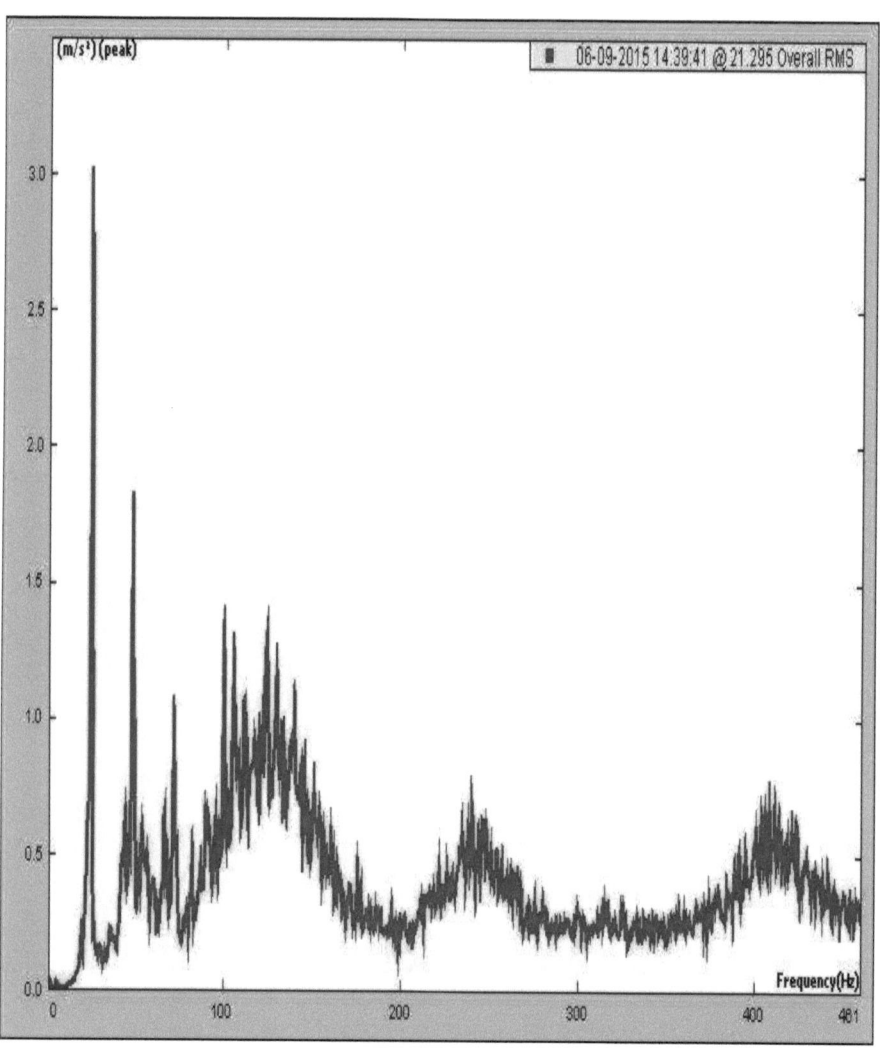

Graph 33: Acceleration Spectrum obtained from FFT analyzer

This spectrum obtained from FFT analyzer gives the acceleration spectrum at 500 RPM with Phasing arrangement. This Graph is plotted between frequency and acceleration. The natural frequency for this spectrum is 19 Hz. And that of maximum acceleration is 3.002 m/s^2.

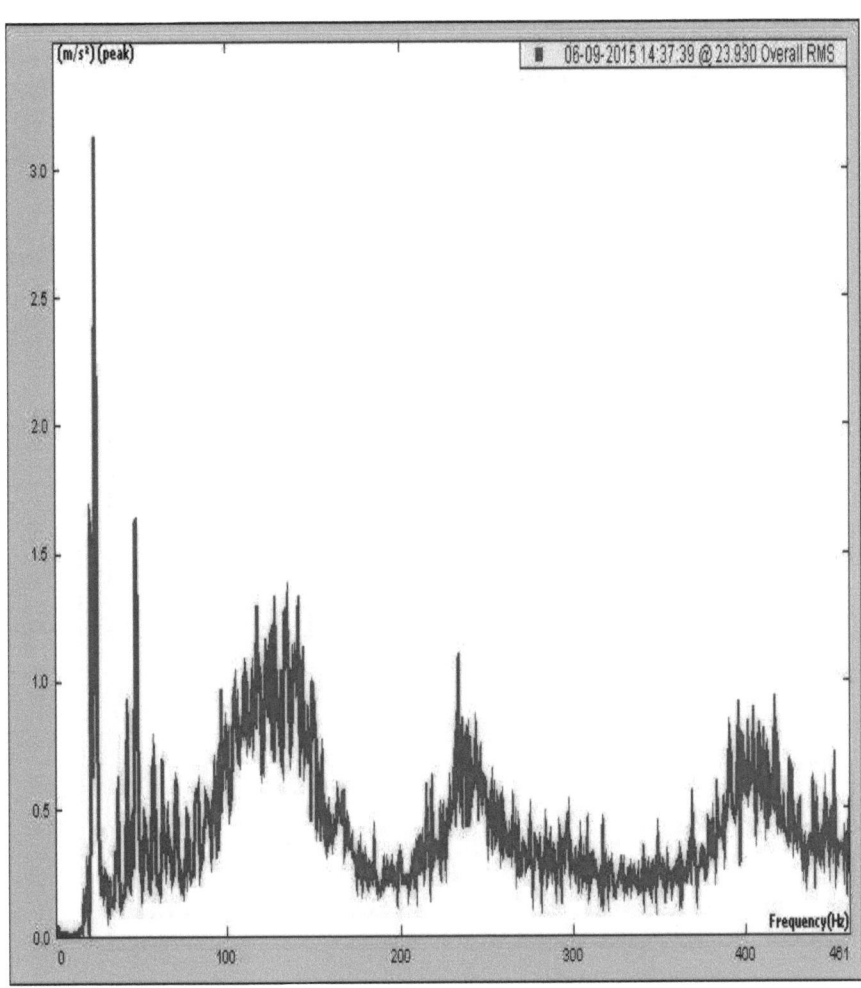

Graph 34: Acceleration Spectrum obtained from FFT analyzer

This spectrum obtained from FFT analyzer gives the acceleration spectrum at 500 RPM without Phasing arrangement. This Graph is plotted between frequency and acceleration. The natural frequency for this spectrum is 21 Hz. And that of maximum acceleration is 3.157 m/s^2.

Table 18: Maximum Acceleration values of single PGT, with and without phasing arrangement obtained from FFT Analyzer

RPM	Maximum Acceleration (m/s^2)		
	Single PGT	With Phasing	Without Phasing
100	1.66	1.16	1.67
200	1.88	1.86	1.91
300	2.11	1.18	2.12
400	2.62	2.61	2.63
500	3.13	3.00	3.15

Summary

This acceleration spectrums gives the peak value of accelerations for single PGT, phasing and without phasing arrangement for various speeds of motor. This conclude that the acceleration of phasing arrangement is reduces as compared to without phasing and single PGT arrangement for the various speeds.

5.5 Displacement spectrums

The displacement spectrums obtained from FFT analyzers for with and without phasing arrangement at various speeds are as follows. The graph 35 to graph 49 gives displacement spectrum as shown in figure.

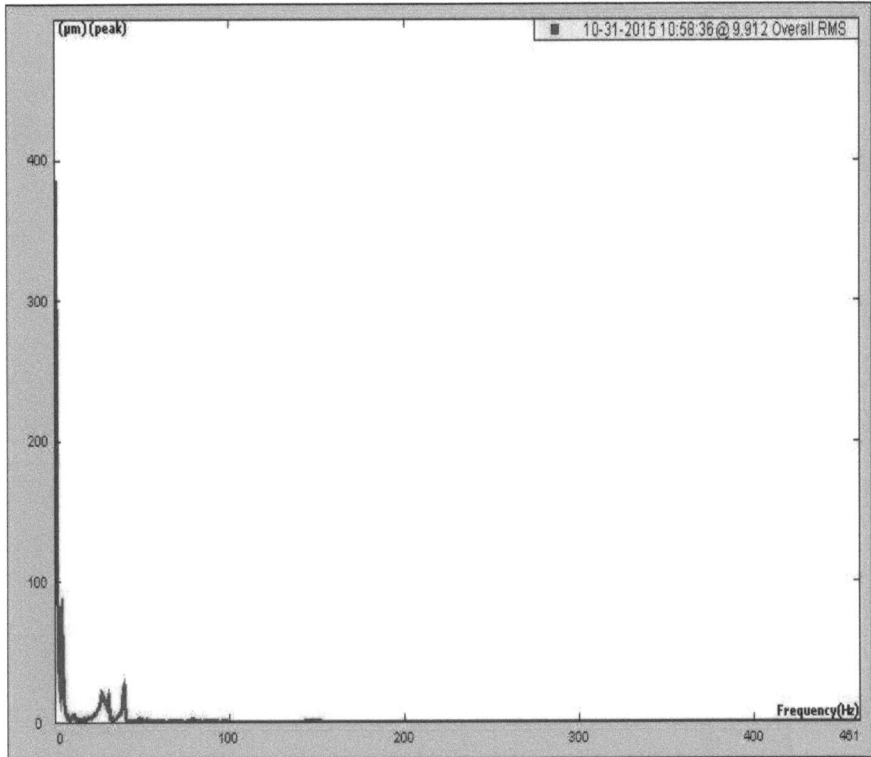

Graph 35: The displacement spectrum obtained from FFT analyzer

This spectrum obtained from FFT analyzer gives the Displacement spectrum at 100 RPM single PGT arrangement. This Graph is plotted between frequency and Displacement. The natural frequency for this spectrum is 40 Hz. And that of maximum Displacement is 35µm.

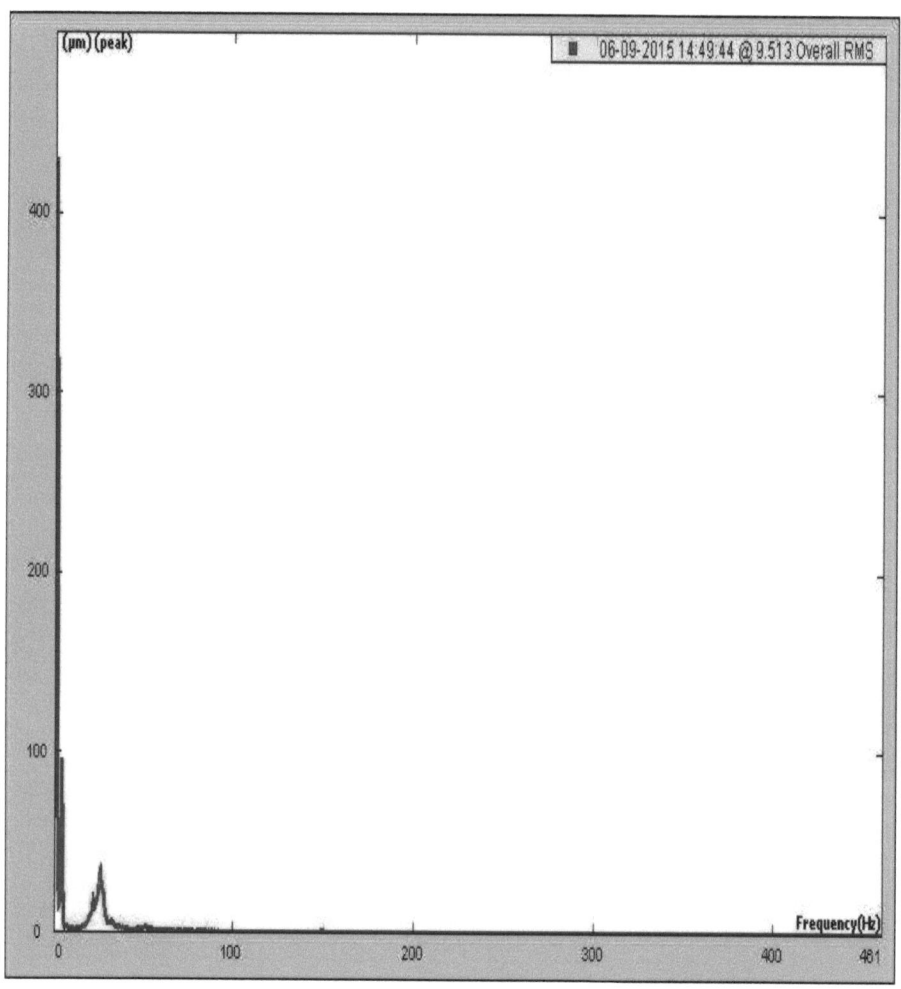

Graph 36: The displacement spectrum obtained from FFT analyzer

This spectrum obtained from FFT analyzer gives the Displacement spectrum at 100 RPM with Phasing arrangement. This Graph is plotted between frequency and Displacement. The natural frequency for this spectrum is 25 Hz. And that of maximum Displacement is 27μm.

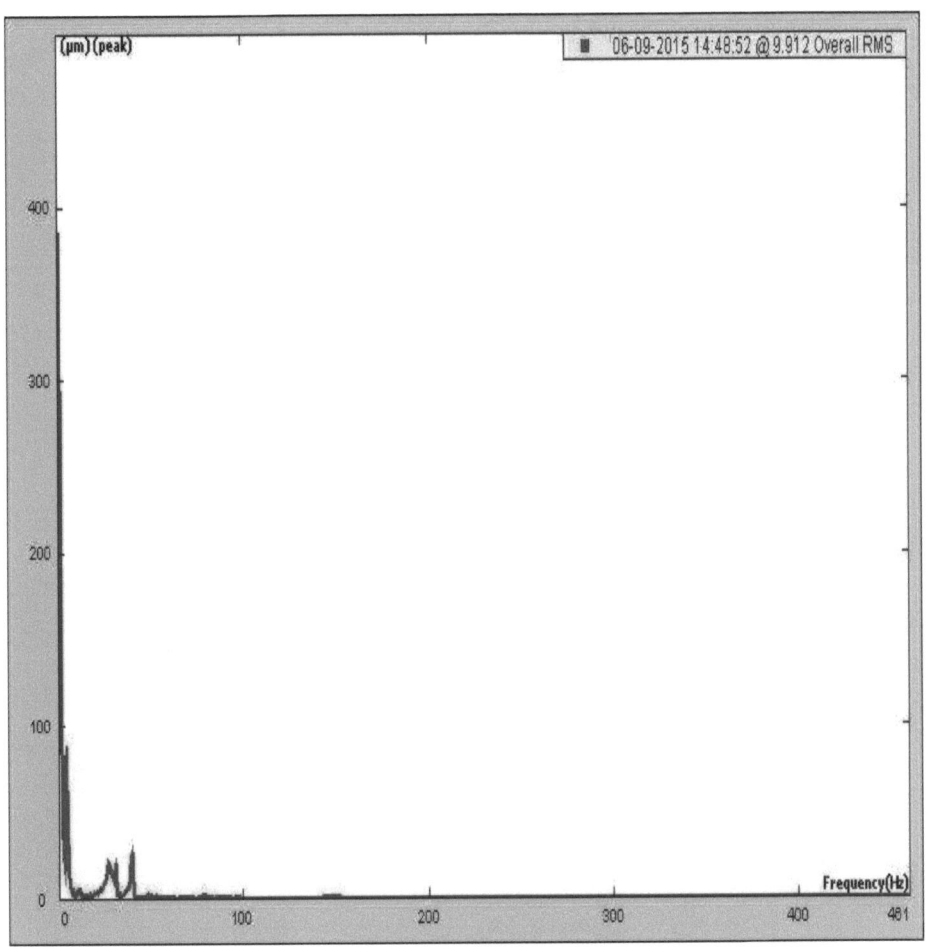

Graph 37: The displacement spectrum obtained from FFT analyzer

This spectrum obtained from FFT analyzer gives the Displacement spectrum at 100 RPM without phasing arrangement. This Graph is plotted between frequency and Displacement. The natural frequency for this spectrum is 40 Hz. And that of maximum Displacement is 38μm.

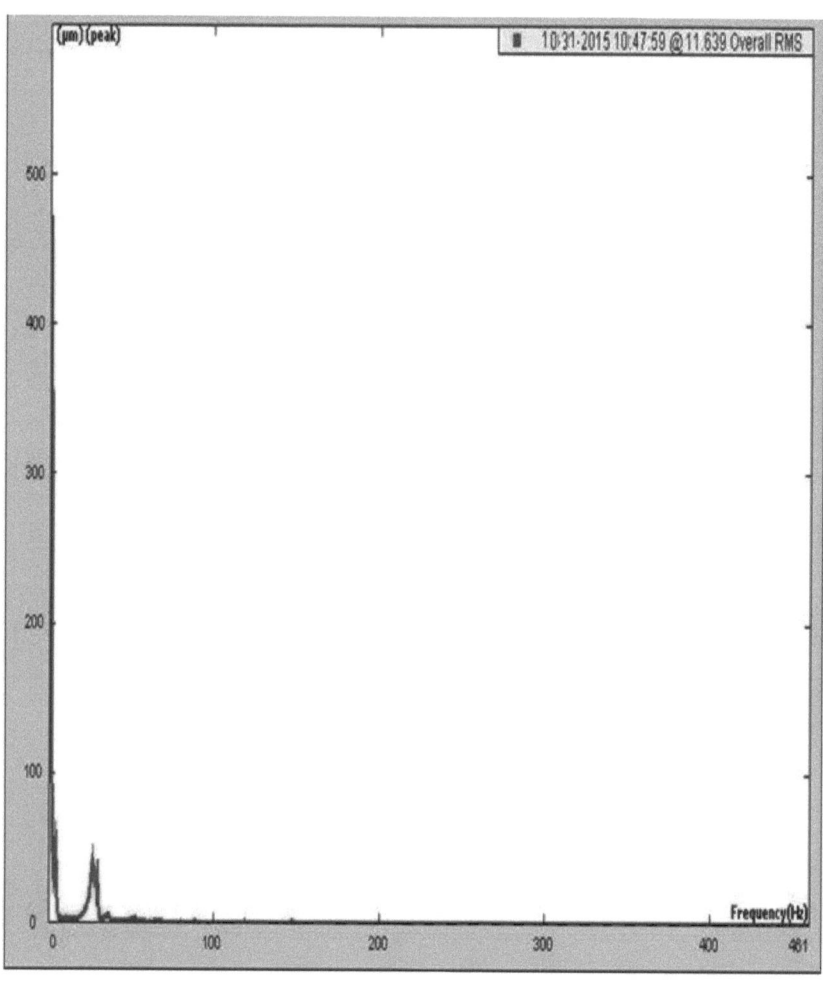

Graph 38: The displacement spectrum obtained from FFT analyzer

This spectrum obtained from FFT analyzer gives the Displacement spectrum at 200 RPM single PGT arrangement. This Graph is plotted between frequency and Displacement. The natural frequency for this spectrum is 25 Hz. And that of maximum Displacement is 40μm.

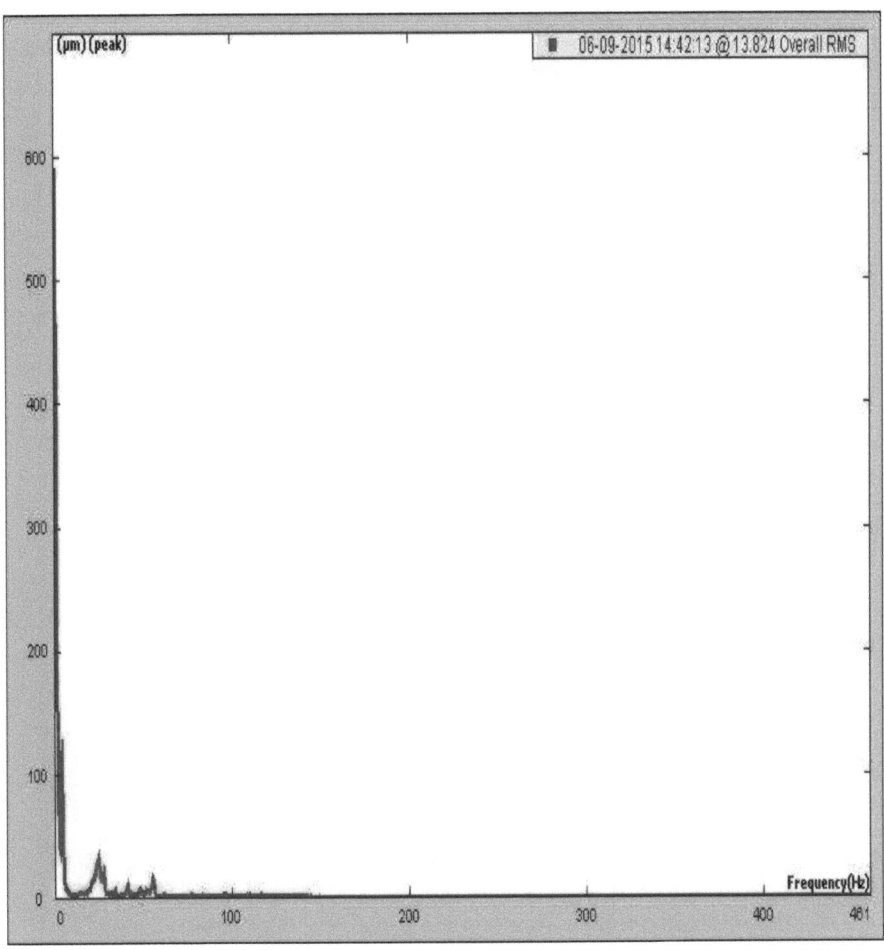

Graph 39: The displacement spectrum obtained from FFT analyzer

This spectrum obtained from FFT analyzer gives the Displacement spectrum at 200 RPM with Phasing arrangement. This Graph is plotted between frequency and Displacement. The natural frequency for this spectrum is 23 Hz. And that of maximum Displacement is 33μm.

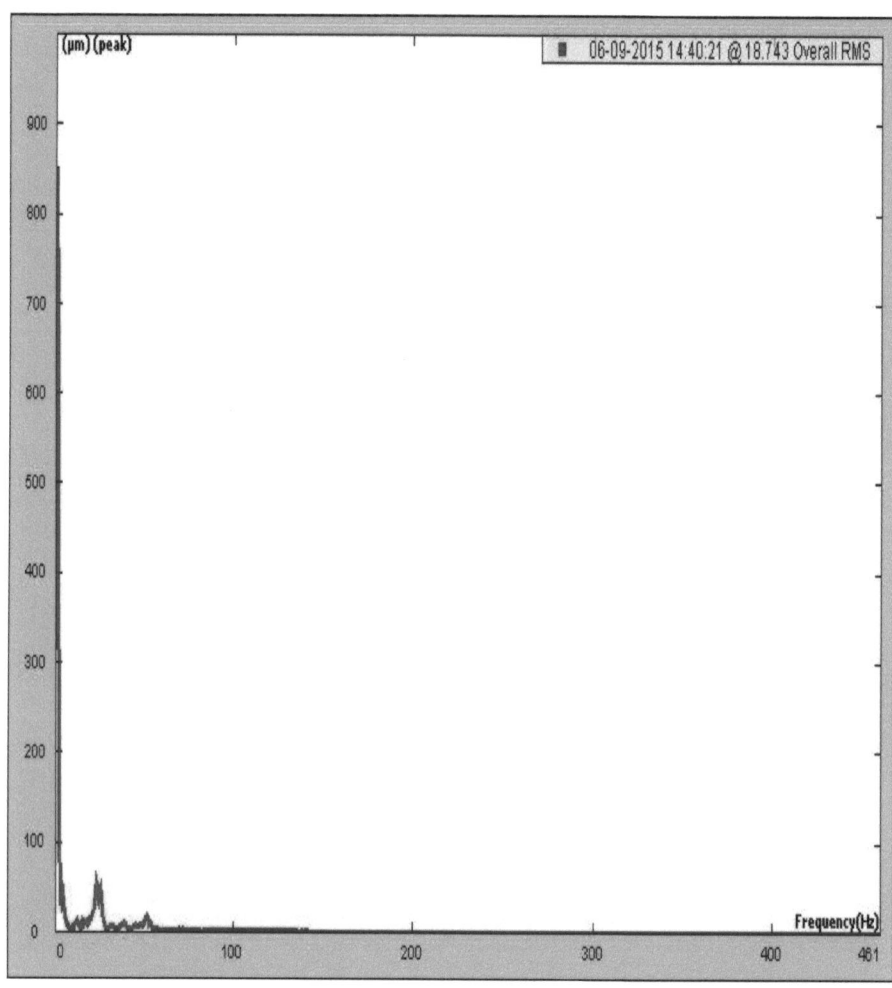

Graph 40: The displacement spectrum obtained from FFT analyzer

This spectrum obtained from FFT analyzer gives the Displacement spectrum at 200 RPM without arrangement. This Graph is plotted between frequency and Displacement. The natural frequency for this spectrum is 25 Hz. And that of maximum Displacement is 60μm.

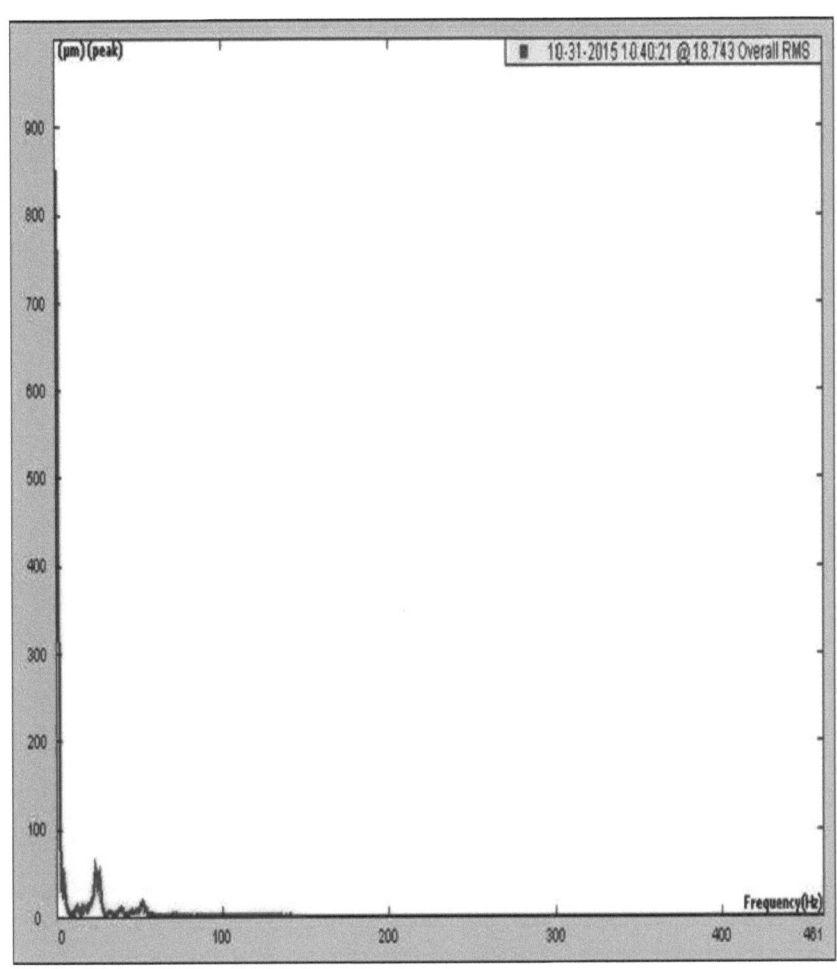

Graph 41: The displacement spectrum obtained from FFT analyzer

This spectrum obtained from FFT analyzer gives the Displacement spectrum at 300 RPM single PGT arrangement. This Graph is plotted between frequency and Displacement. The natural frequency for this spectrum is 25 Hz. And that of maximum Displacement is 75μm.

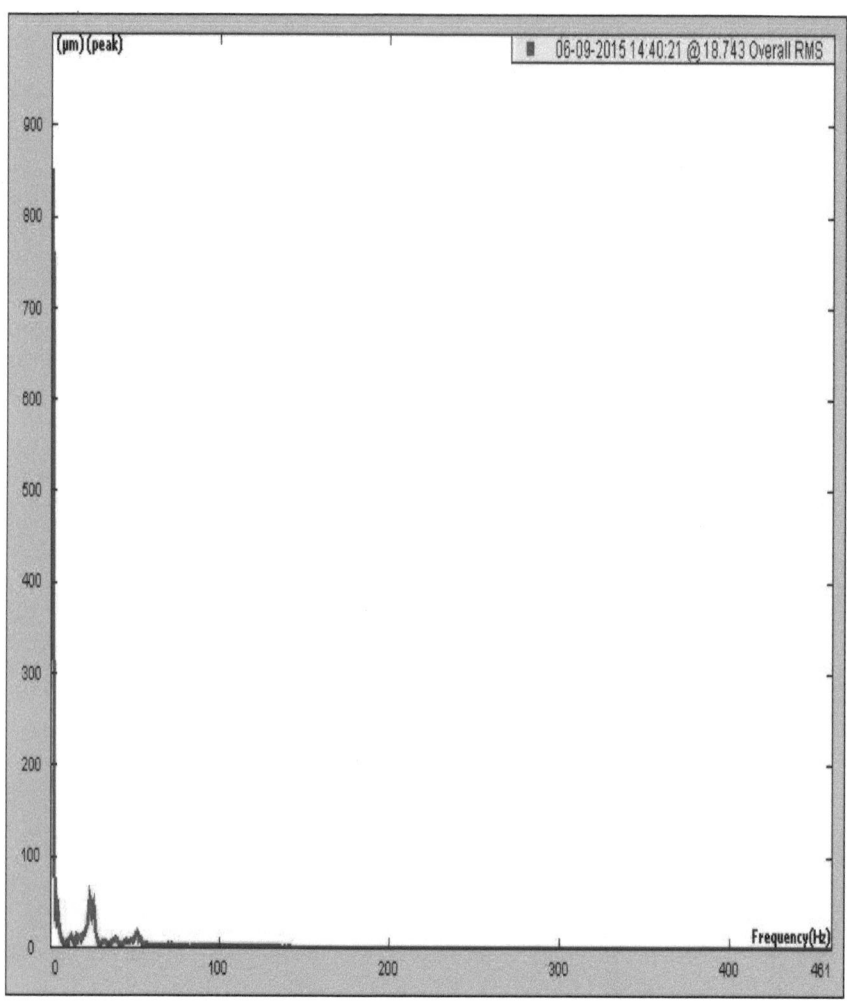

Graph 42: The displacement spectrum obtained from FFT analyzer

This spectrum obtained from FFT analyzer gives the Displacement spectrum at 300 RPM with phasing arrangement. This Graph is plotted between frequency and Displacement. The natural frequency for this spectrum is 23 Hz. And that of maximum Displacement is 65μm.

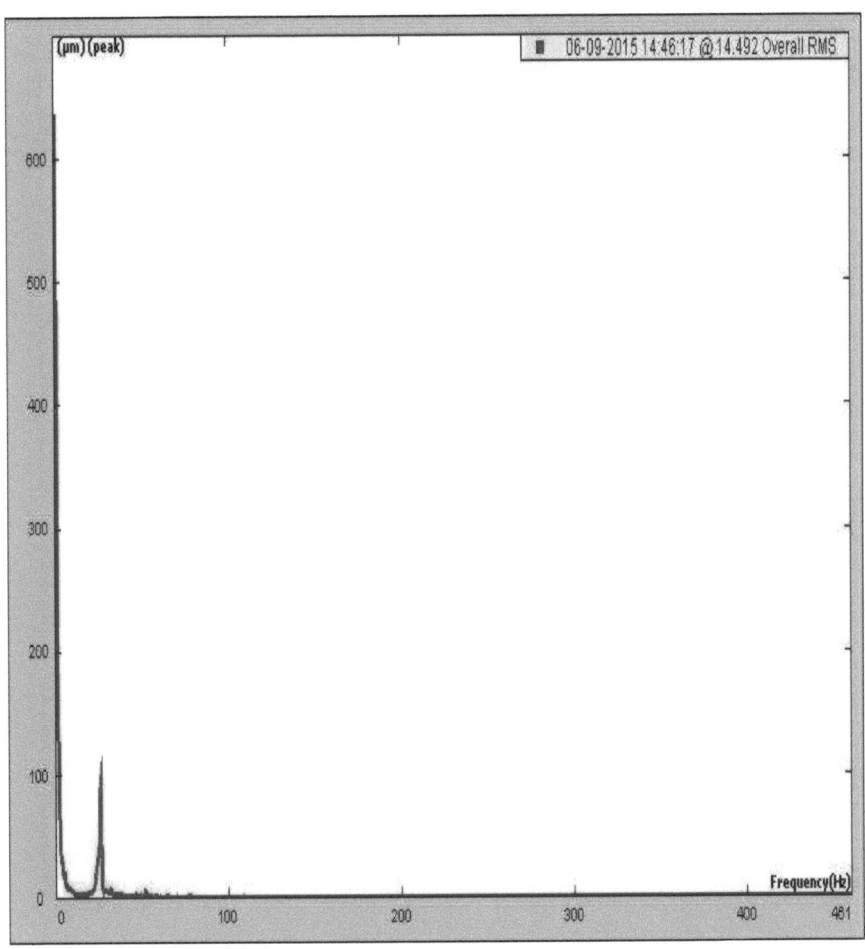

Graph 43: The displacement spectrum obtained from FFT analyzer

This spectrum obtained from FFT analyzer gives the Displacement spectrum at 300 RPM without Phasing arrangement. This Graph is plotted between frequency and Displacement. The natural frequency for this spectrum is 24 Hz. And that of maximum Displacement is 105μm.

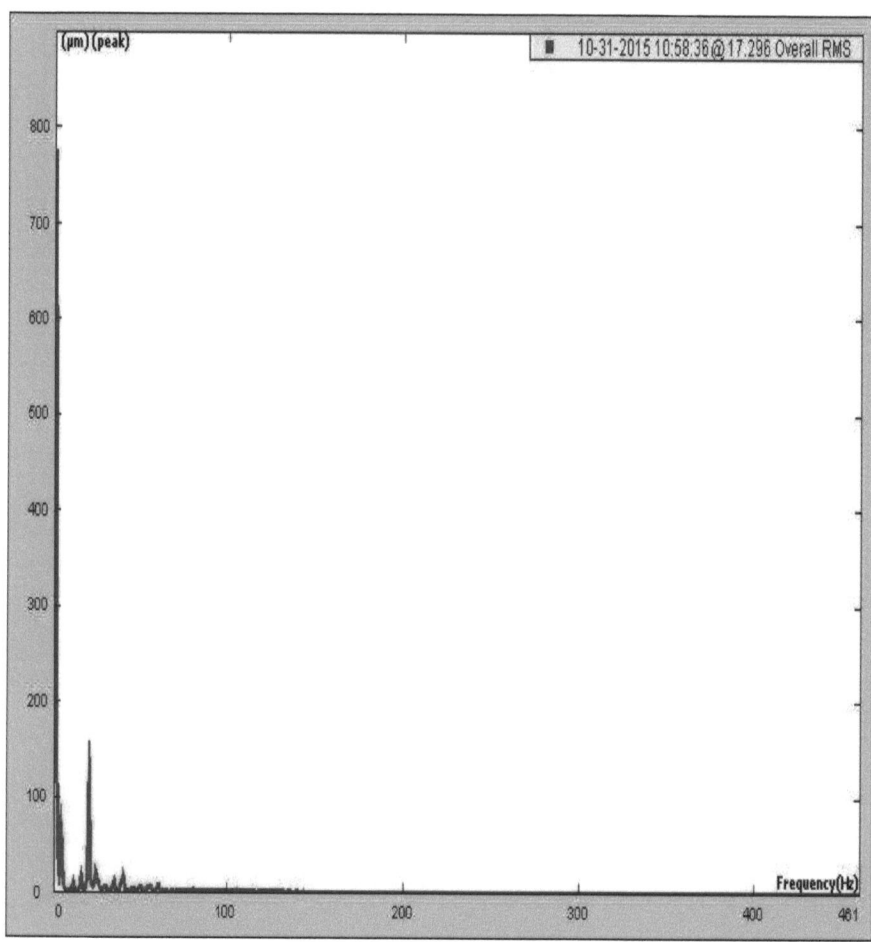

Graph 44: The displacement spectrum obtained from FFT analyzer

This spectrum obtained from FFT analyzer gives the Displacement spectrum at 400 RPM single PGT arrangement. This Graph is plotted between frequency and Displacement. The natural frequency for this spectrum is 23 Hz. And that of maximum Displacement is 155μm.

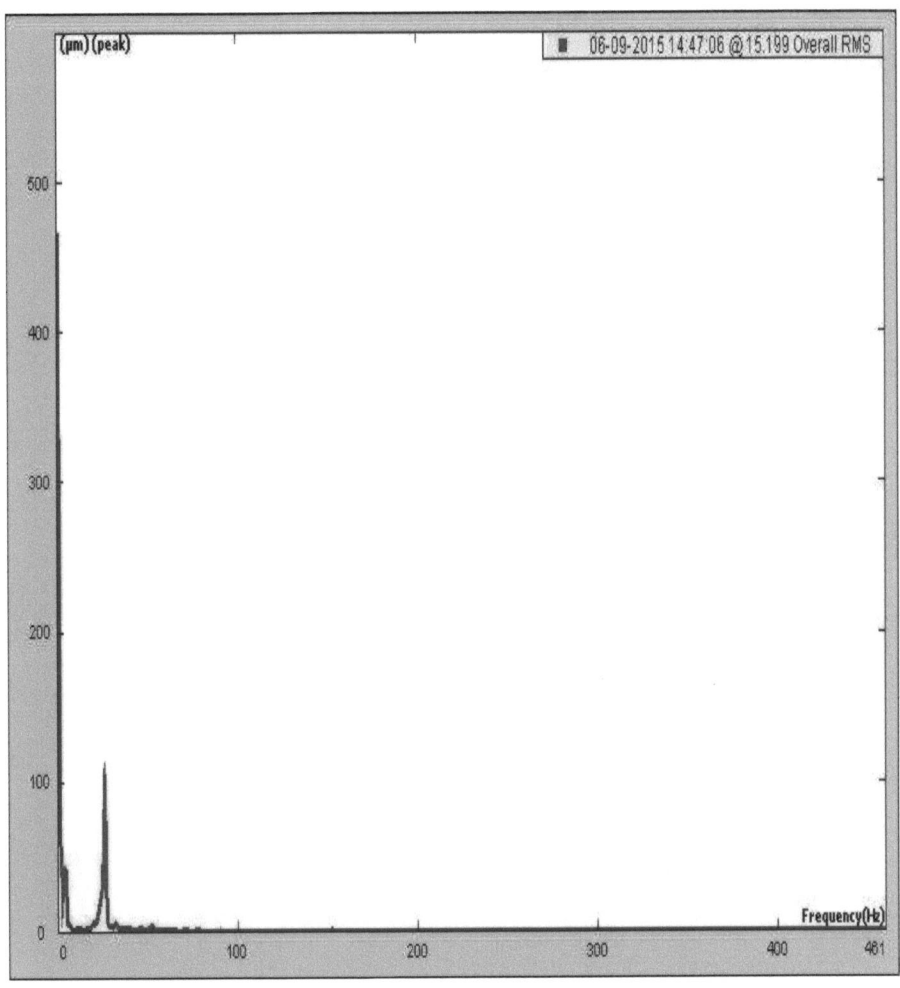

Graph 45: The displacement spectrum obtained from FFT analyzer

This spectrum obtained from FFT analyzer gives the Displacement spectrum at 400 RPM with Phasing arrangement. This Graph is plotted between frequency and Displacement. The natural frequency for this spectrum is 26 Hz. And that of maximum Displacement 138μm.

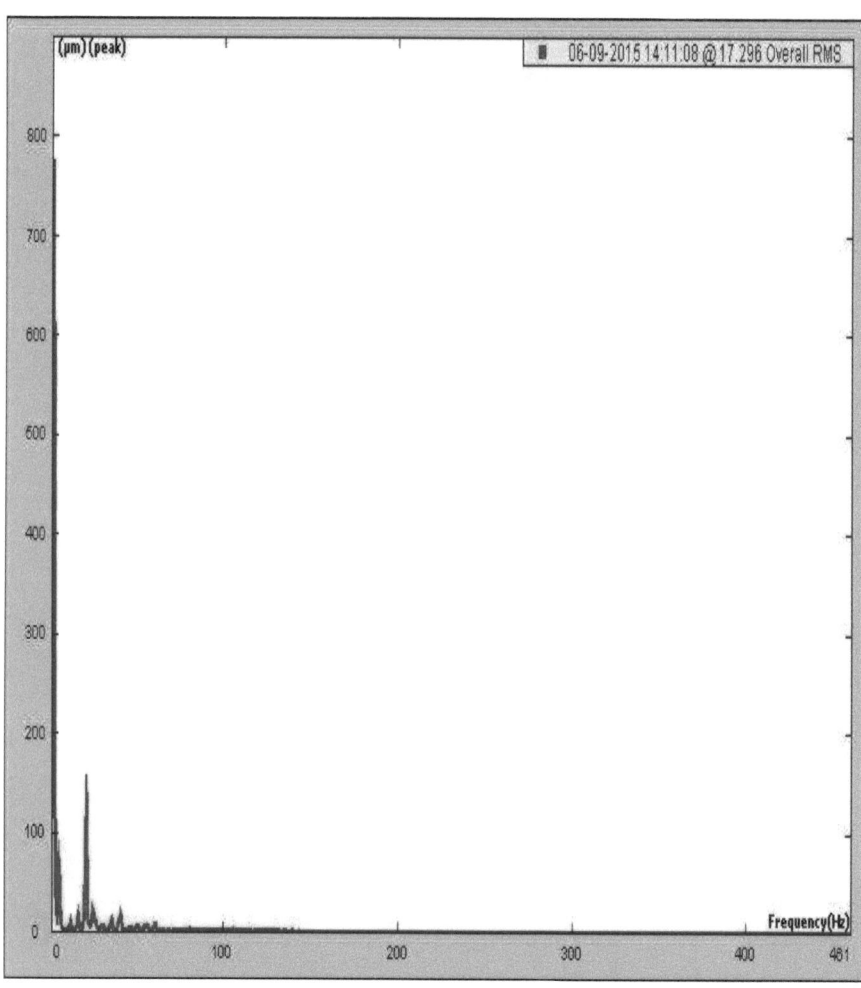

Graph 46: The displacement spectrum obtained from FFT analyzer

This spectrum obtained from FFT analyzer gives the Displacement spectrum at 400 RPM without Phasing arrangement. This Graph is plotted between frequency and Displacement. The natural frequency for this spectrum is 25 Hz. And that of maximum Displacement is 160μm.

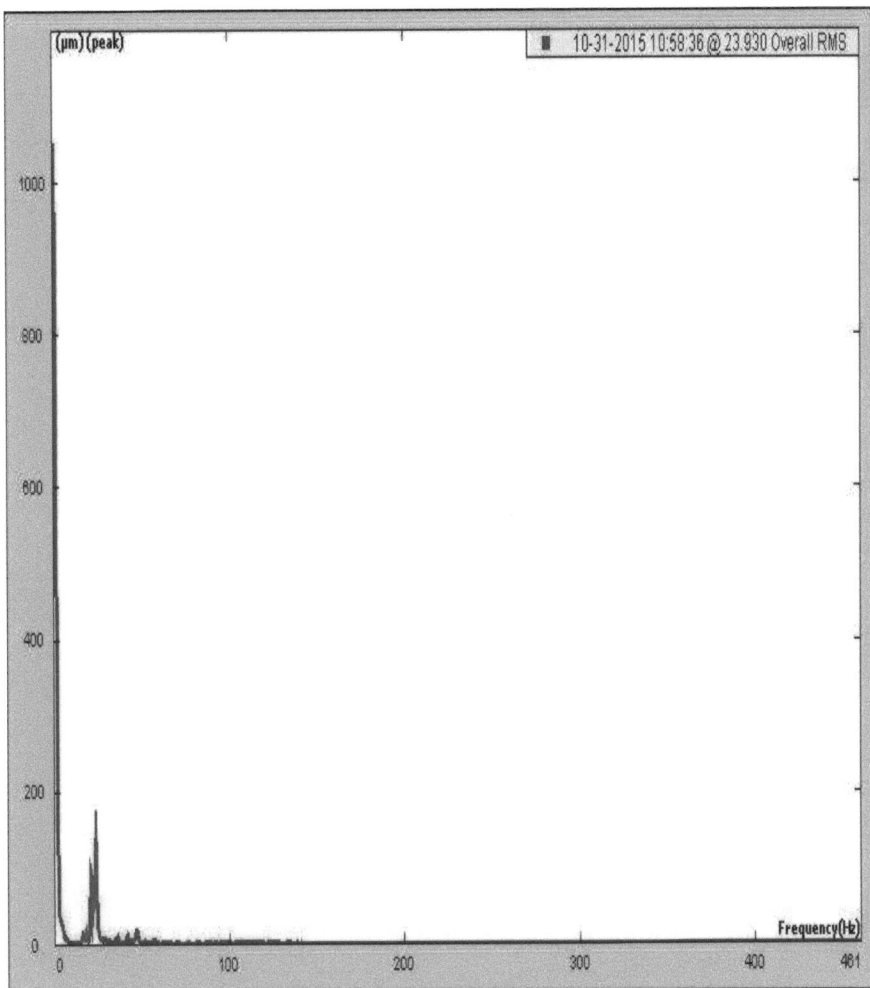

Graph 47: The displacement spectrum obtained from FFT analyzer

This spectrum obtained from FFT analyzer gives the Displacement spectrum at 500 RPM single PGT arrangement. This Graph is plotted between frequency and Displacement. The natural frequency for this spectrum is 22 Hz. And that of maximum Displacement is 180 μm.

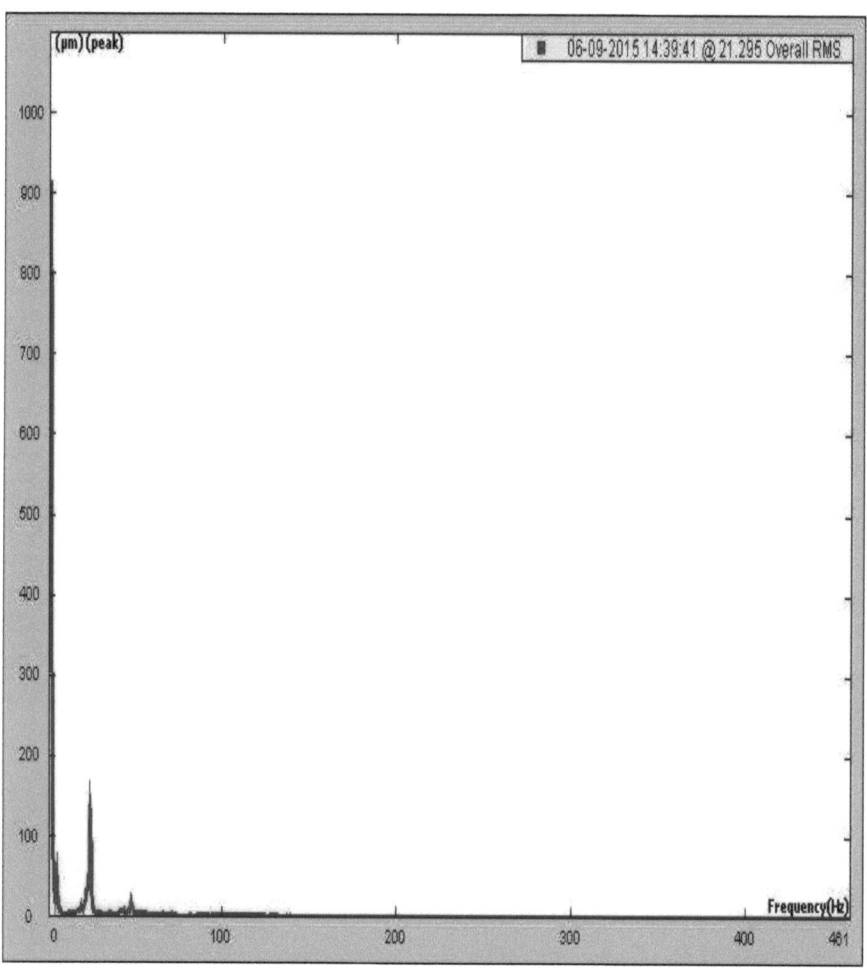

Graph 48: The displacement spectrum obtained from FFT analyzer

This spectrum obtained from FFT analyzer gives the Displacement spectrum at 500 RPM with Phasing arrangement. This Graph is plotted between frequency and Displacement. The natural frequency for this spectrum is 23 Hz. And that of maximum Displacement is 170μm.

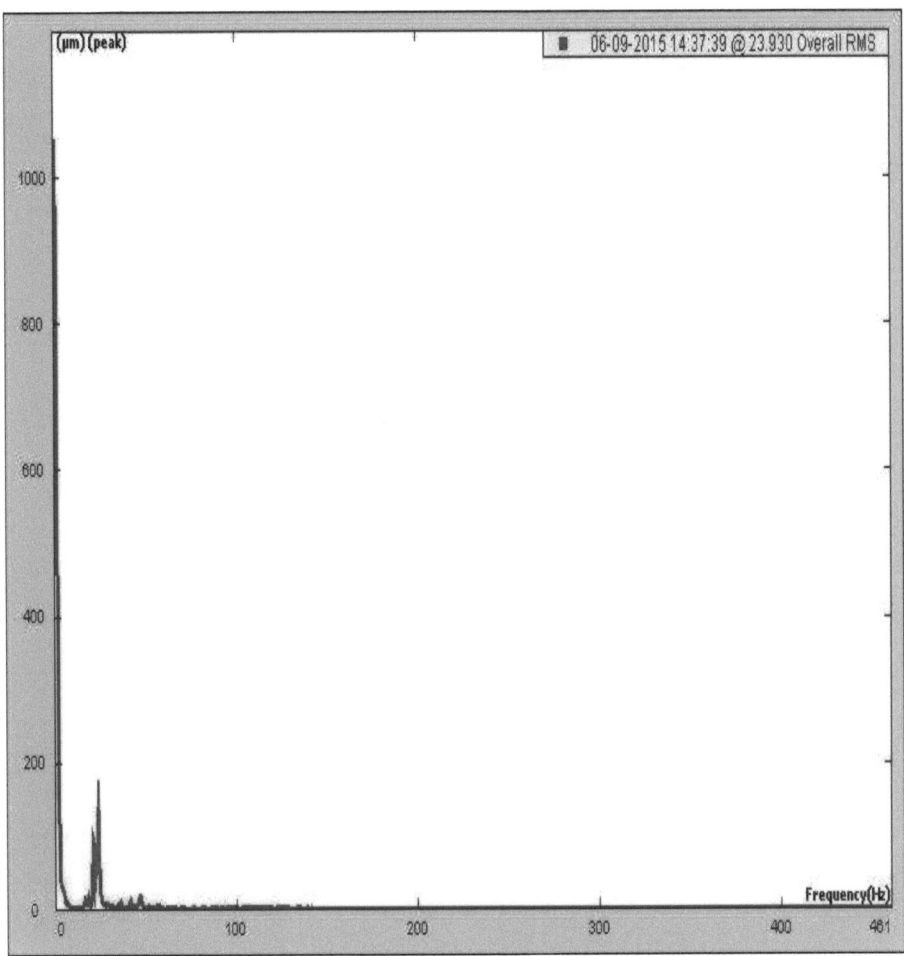

Graph 49: The displacement spectrum obtained from FFT analyzer

This spectrum obtained from FFT analyzer gives the Displacement spectrum at 500 RPM without Phasing arrangement. This Graph is plotted between frequency and Displacement. The natural frequency for this spectrum is 24 Hz. And that of maximum Displacement is 185μm.

Table 19: Maximum Displacement values of single PGT, with and without phasing arrangement obtained from FFT analyzer

RPM	Maximum Displacement (μm)		
	Single PGT	With Phasing	Without Phasing
100	31	27	38
200	40	33	60
300	75	65	105
400	155	138	160
500	180	170	185

Summary

This displacement spectrums gives the peak value of displacements for single PGT, phasing and without phasing arrangement for various speeds of motor. This conclude that the displacement of phasing arrangement is reduces as compared to without phasing arrangement & single PGT arrangement for the various speeds.

5.6 Velocity Spectrums

The velocity spectrums obtained from FFT analyzer for single PGT, with and without phasing arrangement at various speeds as shown in graph 50 to graph 64 at various speeds varying from 0 to 500 RPM.

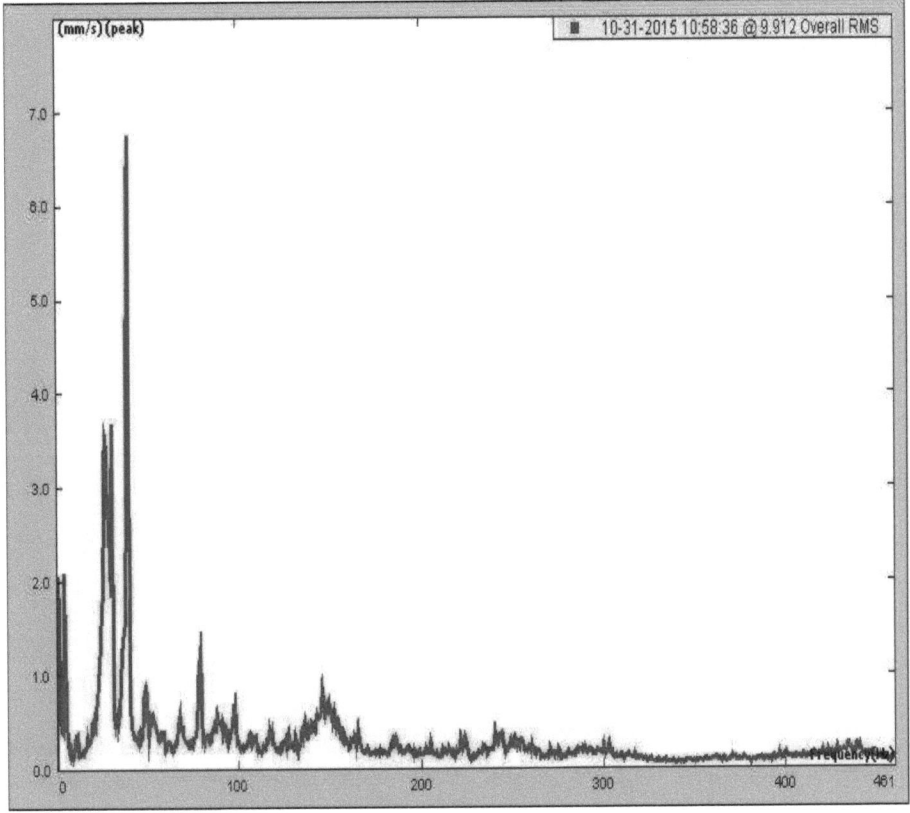

Graph 50: The Velocity spectrum obtained from FFT analyzer

This spectrum obtained from FFT analyzer gives the Velocity spectrum at 100 RPM single PGT arrangement. This Graph is plotted between frequency and Velocity. The natural frequency for this spectrum is 35 Hz. And that of maximum Velocity is 6.6 mm/s.

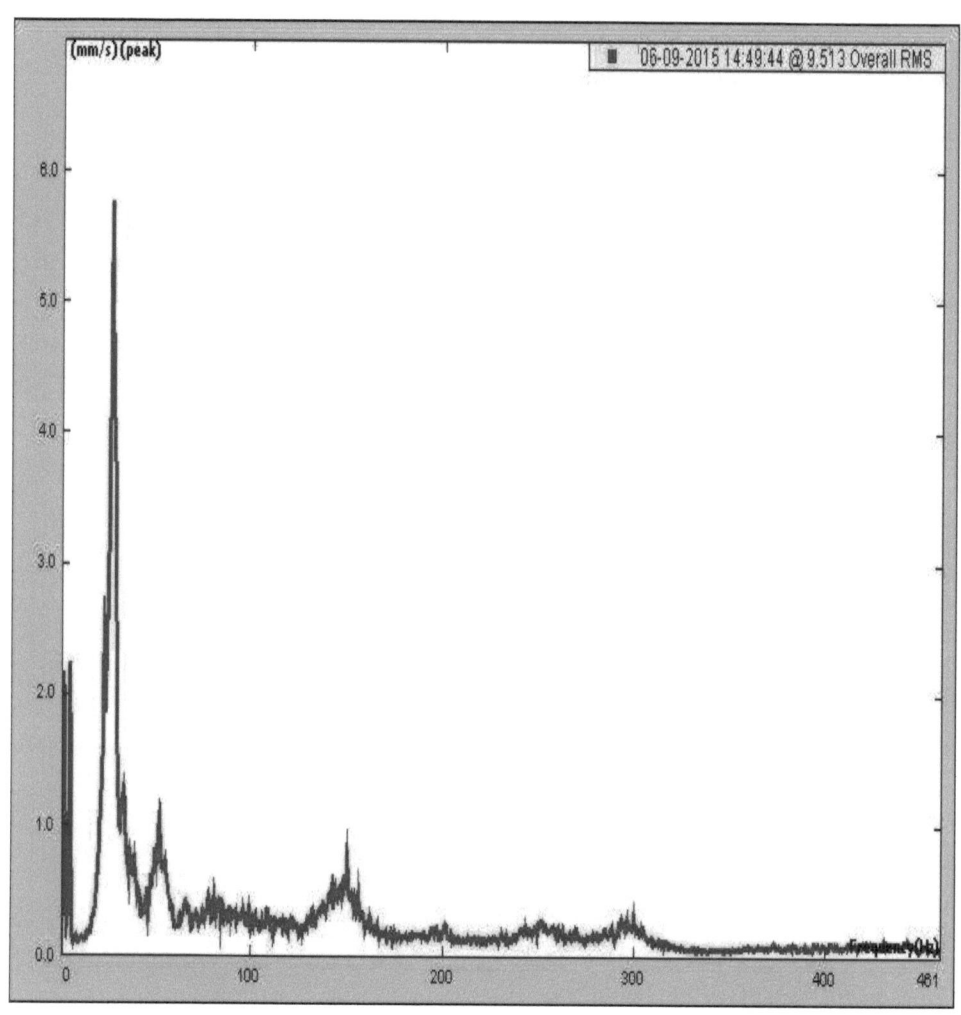

Graph 51: The Velocity spectrum obtained from FFT analyzer

This spectrum obtained from FFT analyzer gives the Velocity spectrum at 100 RPM with Phasing arrangement. This Graph is plotted between frequency and Velocity. The natural frequency for this spectrum is 27 Hz. And that of maximum Velocity is 5.8 mm/s.

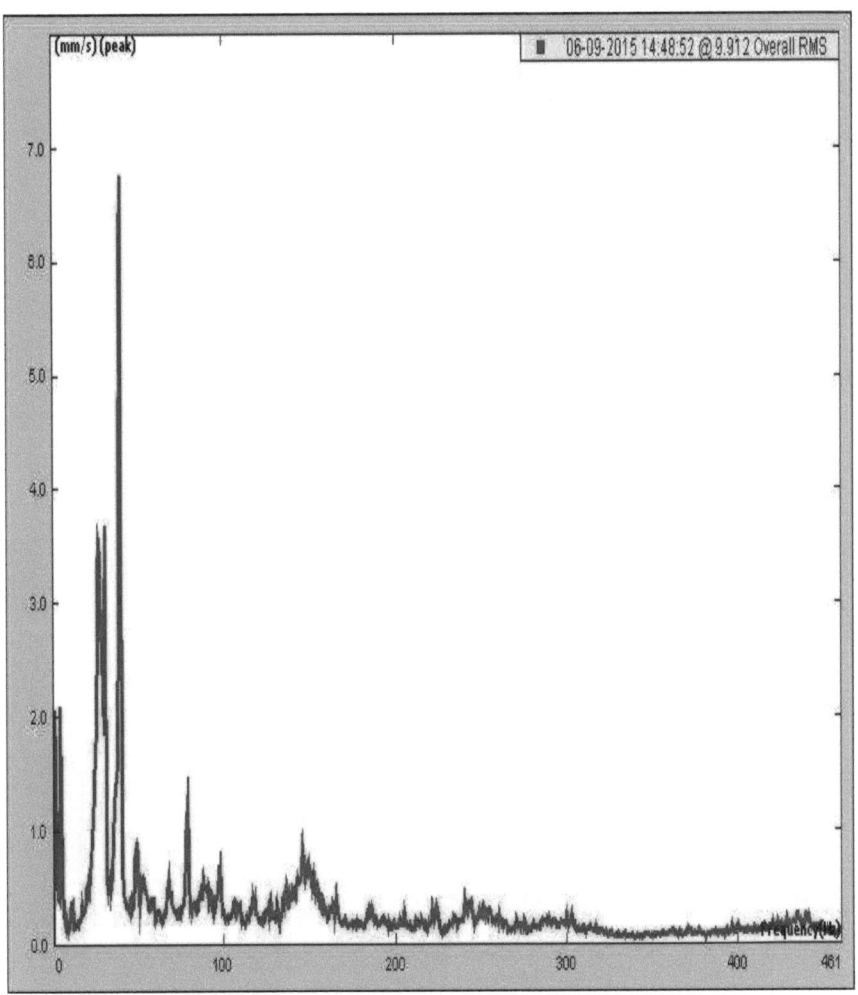

Graph 52: The Velocity spectrum obtained from FFT analyzer

This spectrum obtained from FFT analyzer gives the Velocity spectrum at 100 RPM without Phasing arrangement. This Graph is plotted between frequency and Velocity. The natural frequency for this spectrum is 37 Hz. And that of maximum Velocity is 6.8 mm/s.

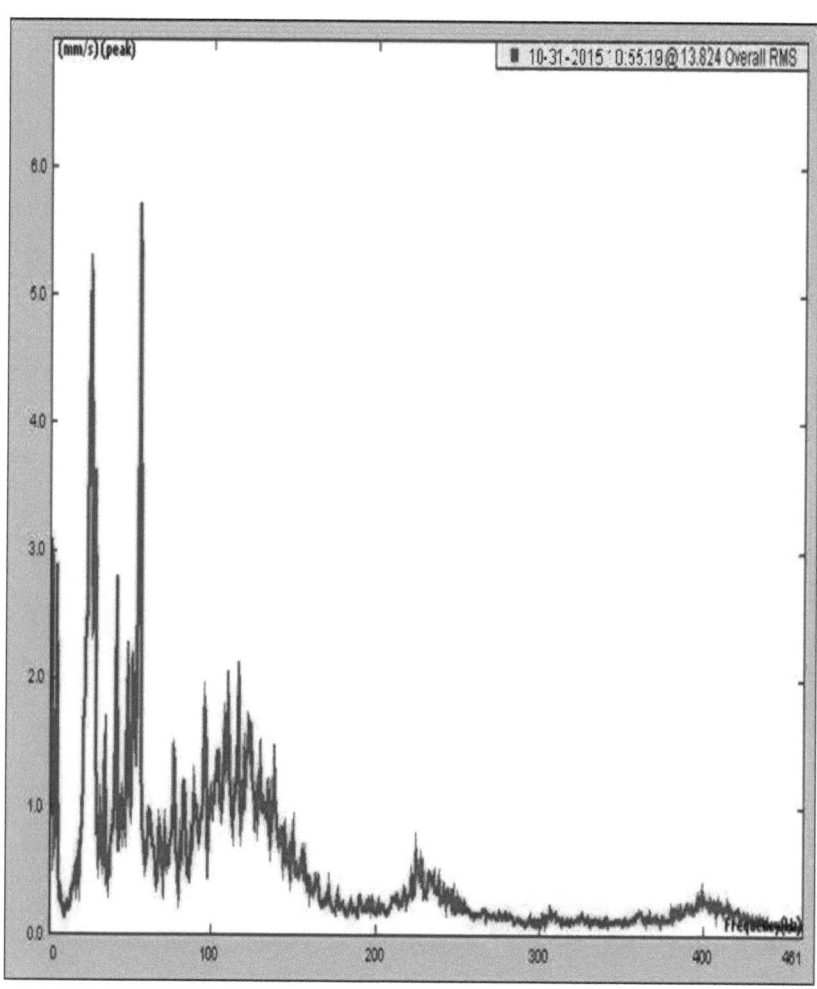

Graph 53: The Velocity spectrum obtained from FFT analyzer

This spectrum obtained from FFT analyzer gives the Velocity spectrum at 200 RPM single phasing arrangement. This Graph is plotted between frequency and Velocity. The natural frequency for this spectrum is 53 Hz. And that of maximum Velocity is 5.7 mm/s.

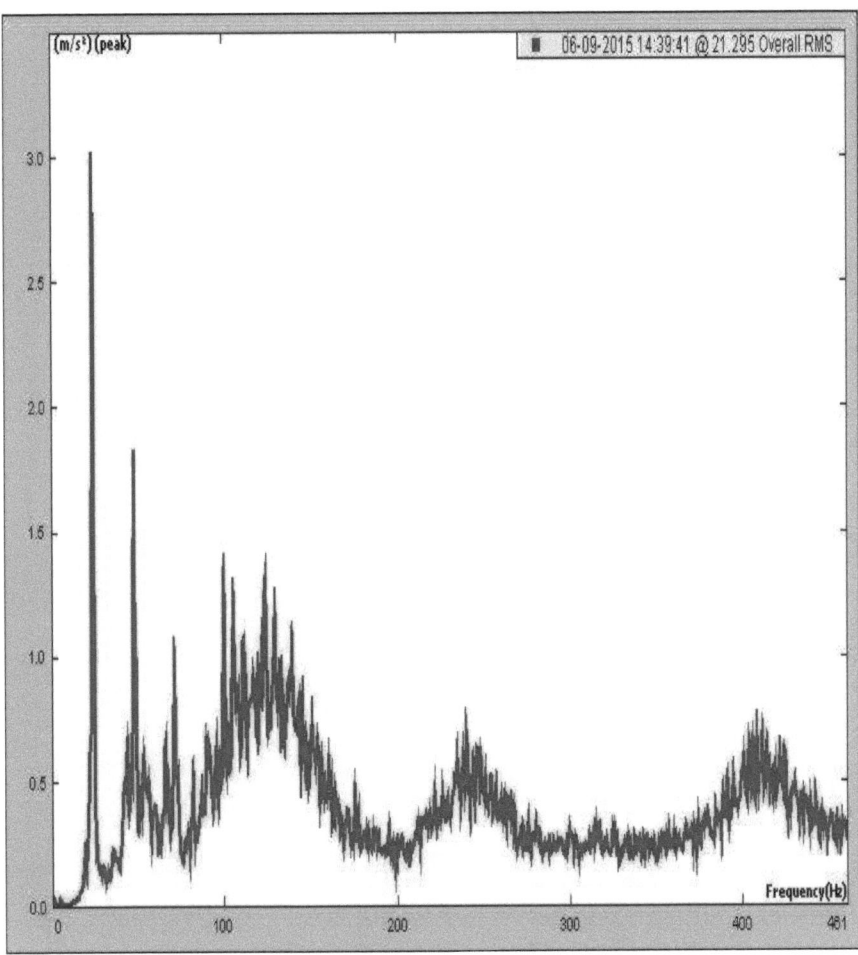

Graph 54: The Velocity spectrum obtained from FFT analyzer

This spectrum obtained from FFT analyzer gives the Velocity spectrum at 200 RPM with Phasing arrangement. This Graph is plotted between frequency and Velocity. The natural frequency for this spectrum is 22 Hz. And that of maximum Velocity is 3.1 mm/s.

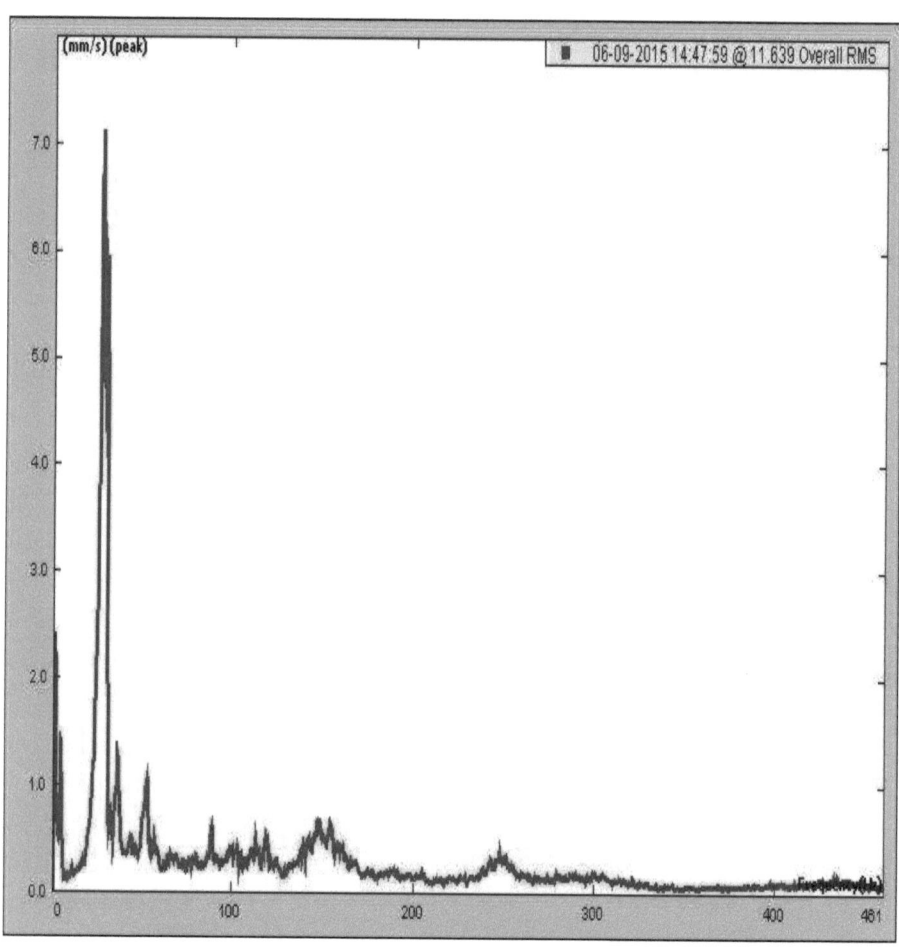

Graph 55: The Velocity spectrum obtained from FFT analyzer

This spectrum obtained from FFT analyzer gives the Velocity spectrum at 200 RPM without Phasing arrangement. This Graph is plotted between frequency and Velocity. The natural frequency for this spectrum is 24 Hz. And that of maximum Velocity is 7.1 mm/s.

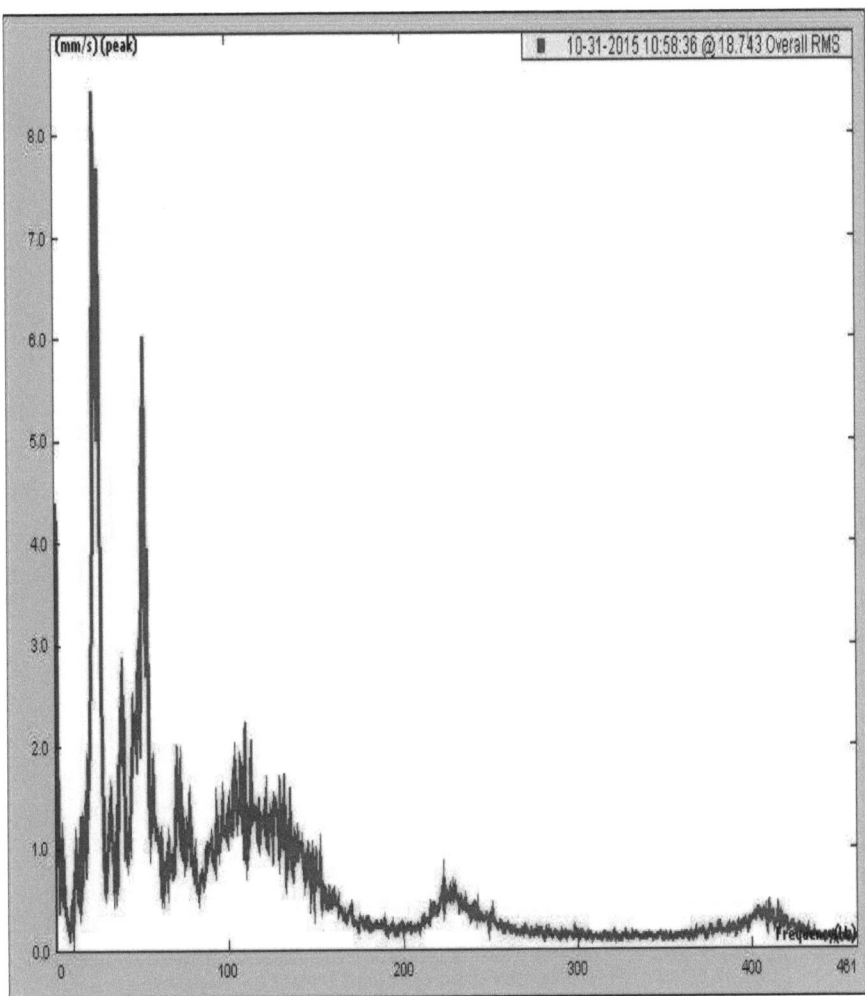

Graph 56: The Velocity spectrum obtained from FFT analyzer

This spectrum obtained from FFT analyzer gives the Velocity spectrum at 300 RPM single PGT arrangement. This Graph is plotted between frequency and Velocity. The natural frequency for this spectrum is 24 Hz. And that of maximum Velocity is 8.3 mm/s.

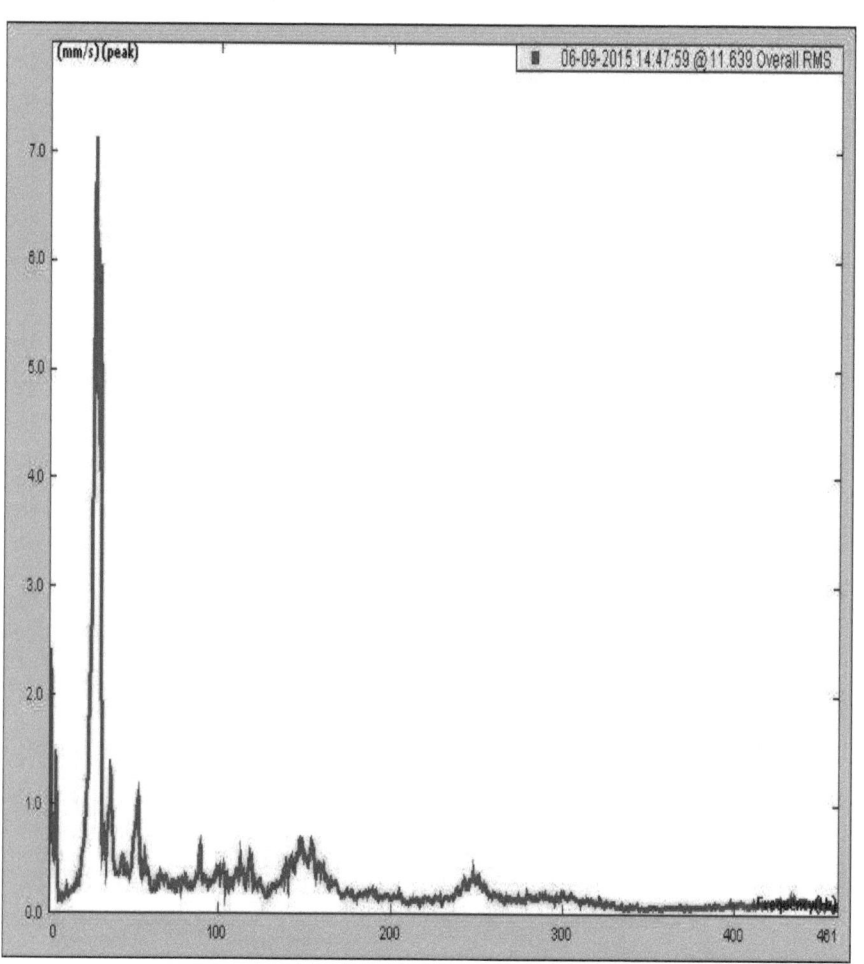

Graph 57: The Velocity spectrum obtained from FFT analyzer

This spectrum obtained from FFT analyzer gives the Velocity spectrum at 300 RPM with phasing arrangement. This Graph is plotted between frequency and Velocity. The natural frequency for this spectrum is 20 Hz. And that of maximum Velocity is 7.2 mm/s.

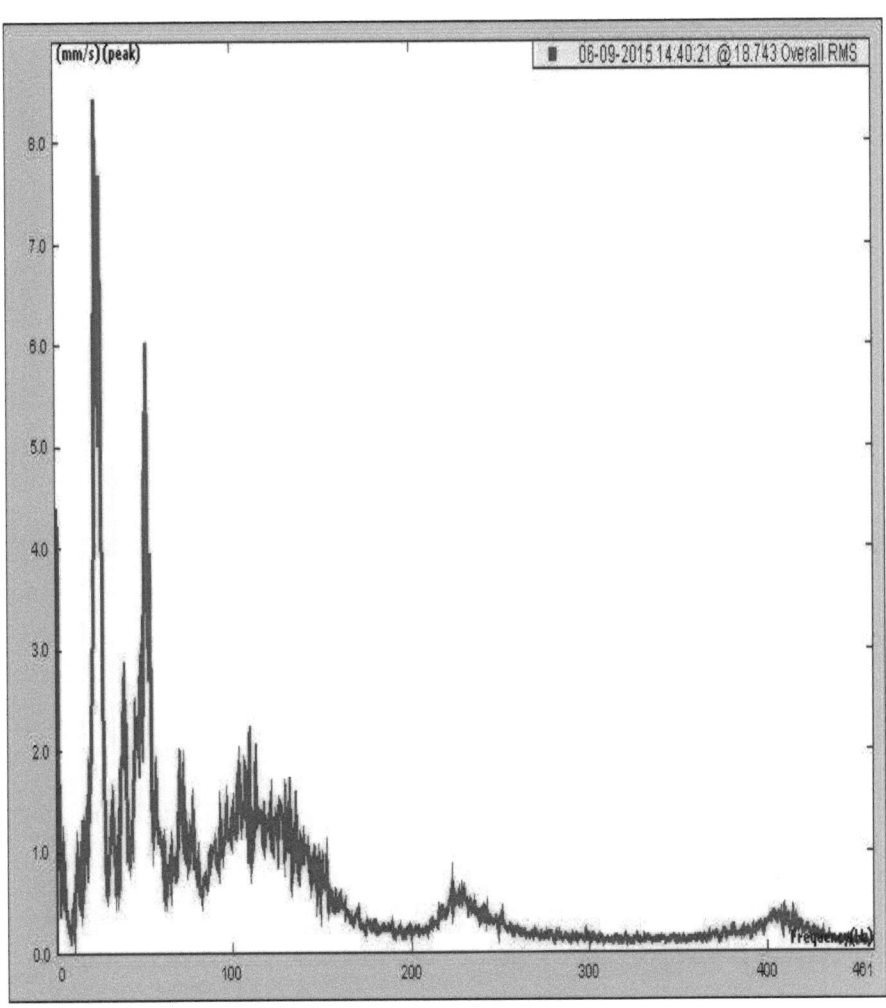

Graph 58: The Velocity spectrum obtained from FFT analyzer

This spectrum obtained from FFT analyzer gives the Velocity spectrum at 300 RPM without Phasing arrangement. This Graph is plotted between frequency and Velocity. The natural frequency for this spectrum is 25 Hz. And that of maximum Velocity is 8.4 mm/s.

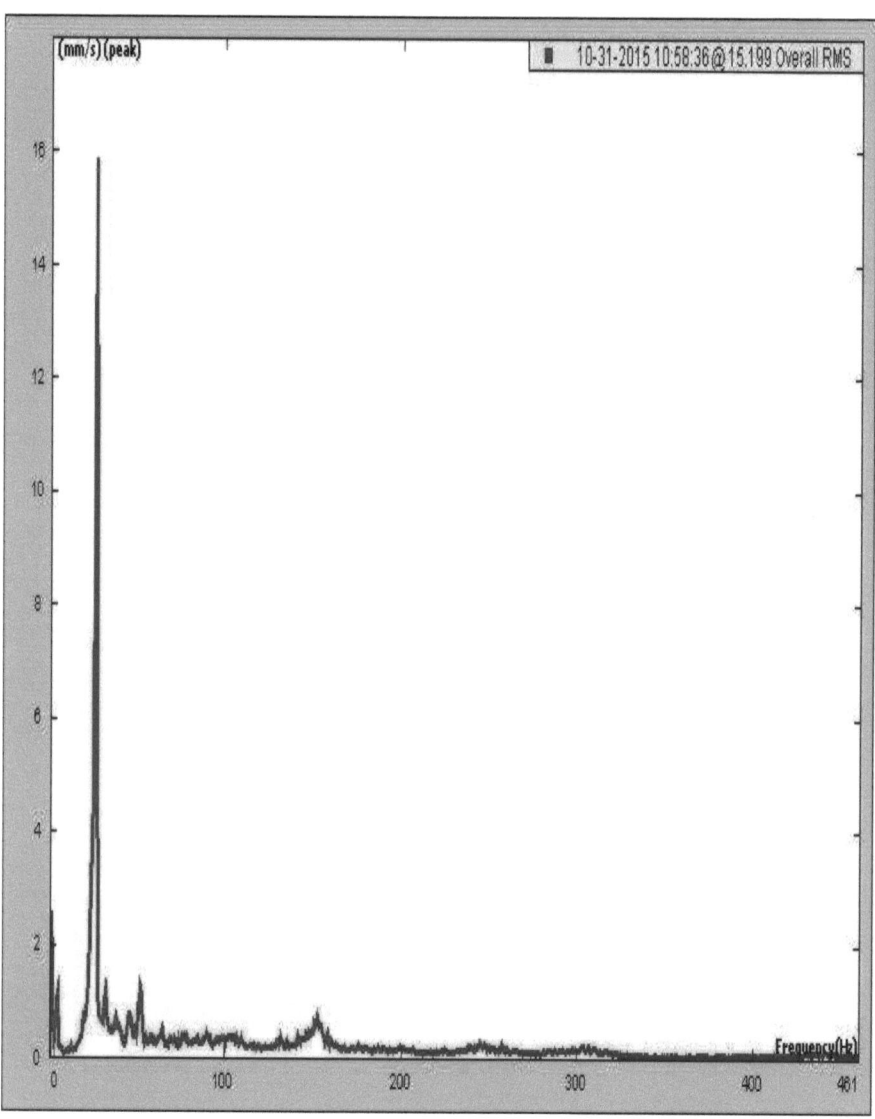

Graph 59: The Velocity spectrum obtained from FFT analyzer

This spectrum obtained from FFT analyzer gives the Velocity spectrum at 400 RPM single PGT arrangement. This Graph is plotted between frequency and Velocity. The natural frequency for this spectrum is 25 Hz. And that of maximum Velocity is 15.9 mm/

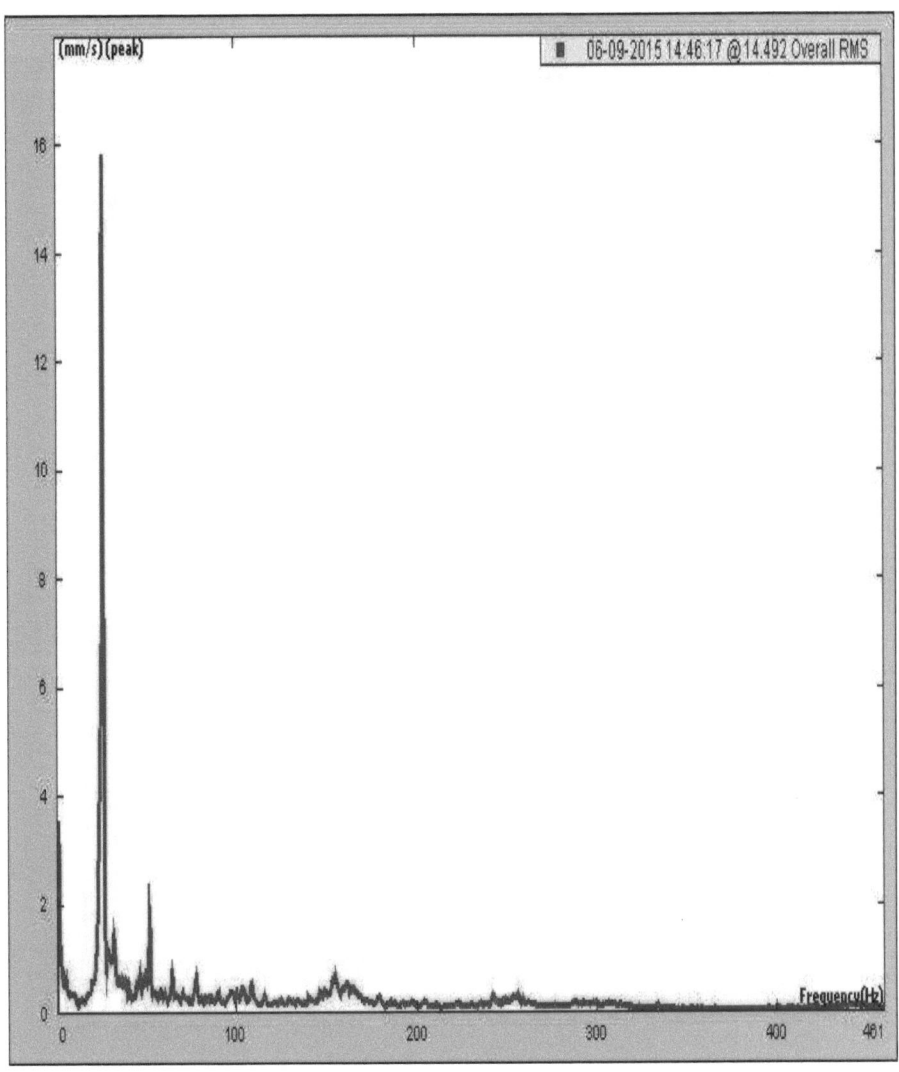

Graph 60: The Velocity spectrum obtained from FFT analyzer

This spectrum obtained from FFT analyzer gives the Velocity spectrum at 400 RPM with Phasing arrangement. This Graph is plotted between frequency and Velocity. The natural frequency for this spectrum is 24 Hz. And that of maximum Velocity is 15.8 mm/s.

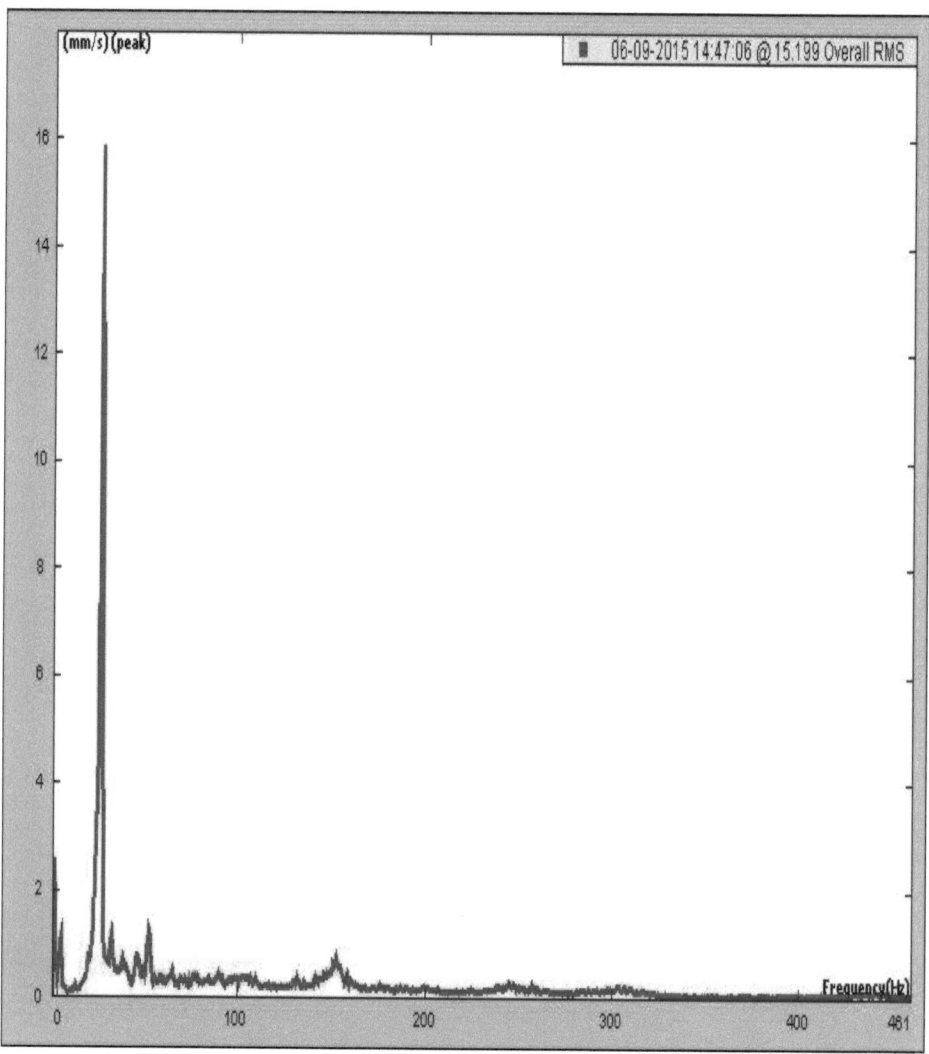

Graph 61: The Velocity spectrum obtained from FFT analyzer

This spectrum obtained from FFT analyzer gives the Velocity spectrum at 400 RPM without phasing arrangement. This Graph is plotted between frequency and Velocity. The natural frequency for this spectrum is 25 Hz. And that of maximum Velocity is 15.95 mm/s.

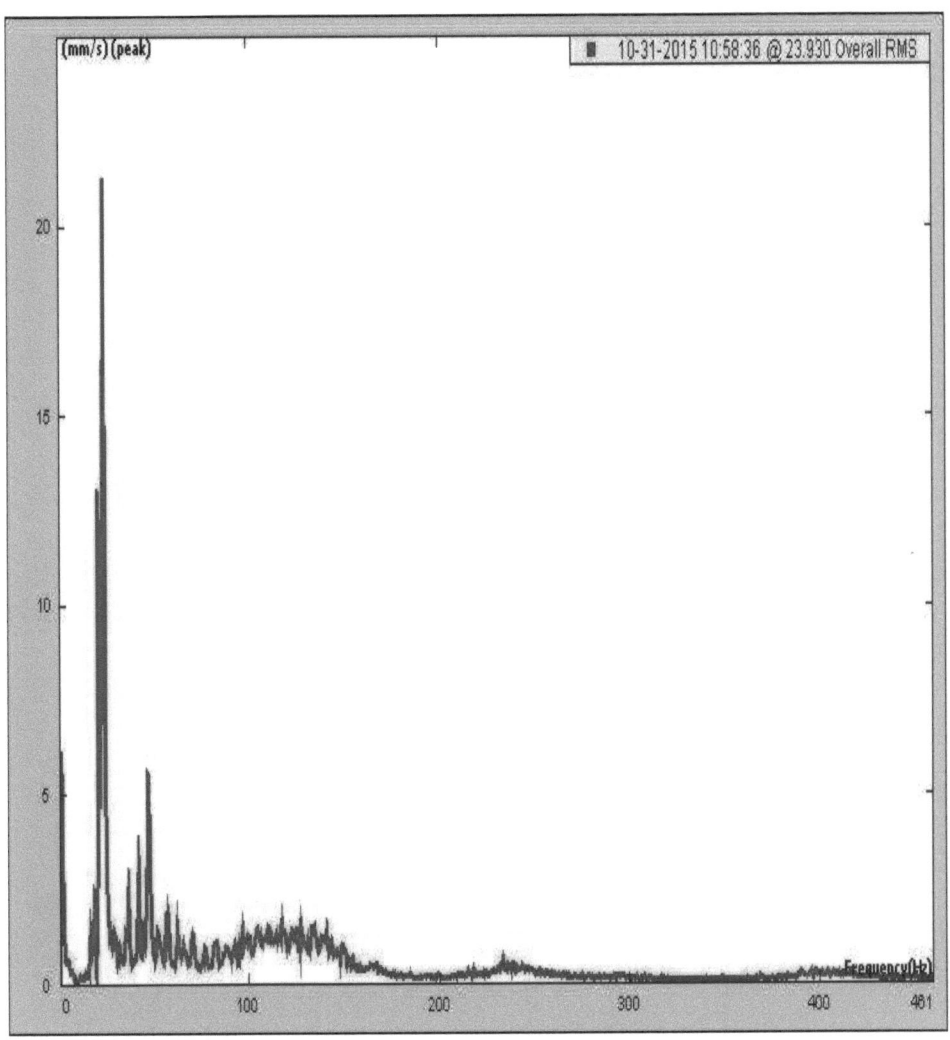

Graph 62: The Velocity spectrum obtained from FFT analyzer

This spectrum obtained from FFT analyzer gives the Velocity spectrum at 500 RPM single PGT arrangement. This Graph is plotted between frequency and Velocity. The natural frequency for this spectrum is 25 Hz. And that of maximum Velocity is 21.5 mm/s.

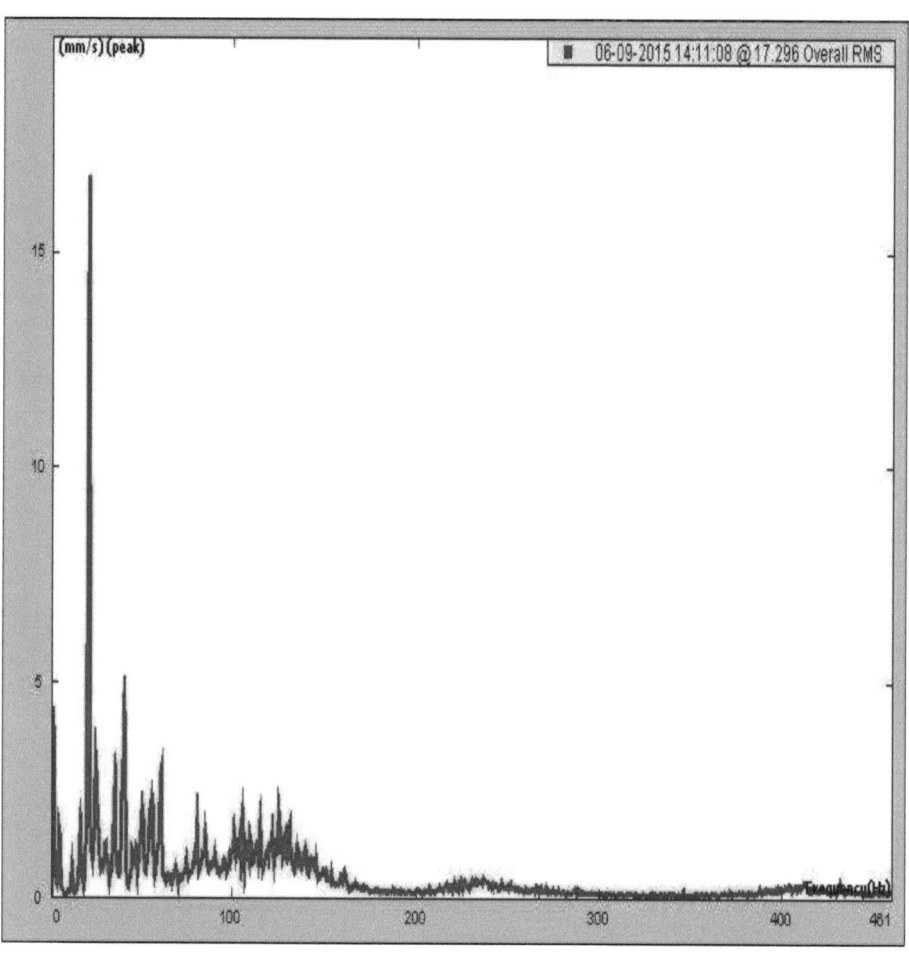

Graph 63: The Velocity spectrum obtained from FFT analyzer

This spectrum obtained from FFT analyzer gives the Velocity spectrum at 500 RPM with Phasing arrangement. This Graph is plotted between frequency and Velocity. The natural frequency for this spectrum is 21 Hz. And that of maximum Velocity is 17.3 mm/s.

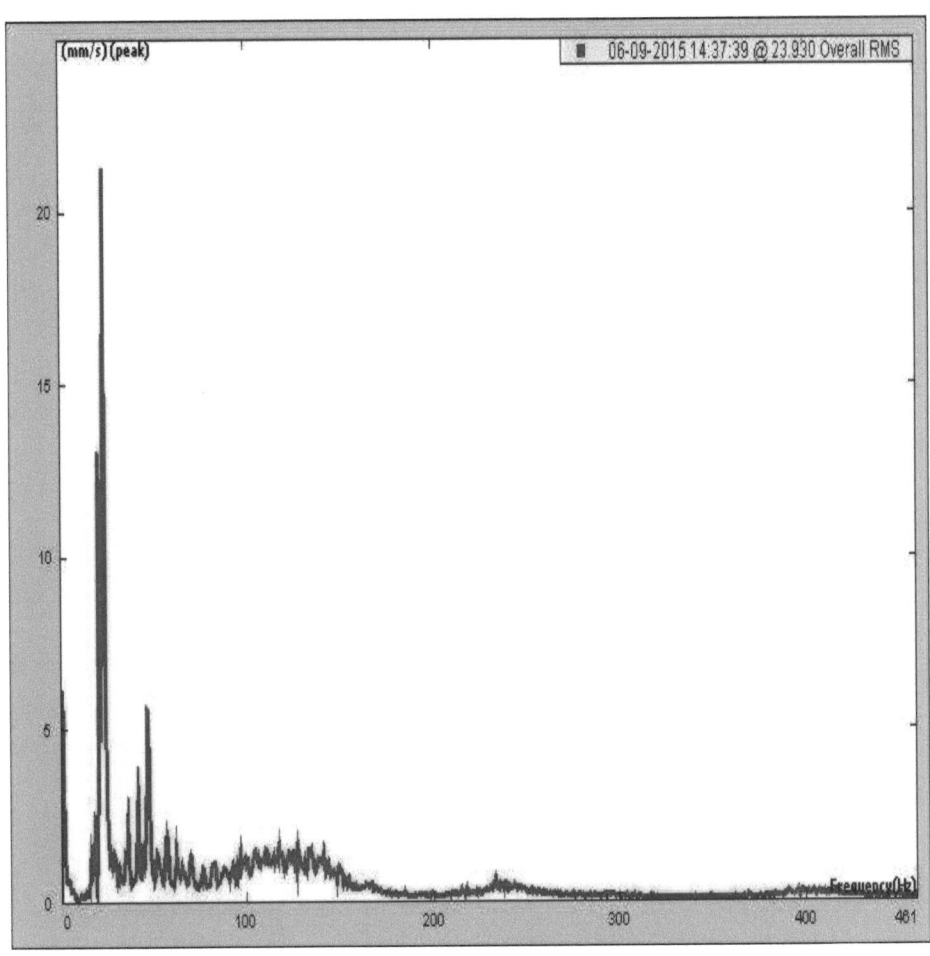

Graph 64: The Velocity spectrum obtained from FFT analyzer

This spectrum obtained from FFT analyzer gives the Velocity spectrum at 500 RPM without Phasing arrangement. This Graph is plotted between frequency and Velocity. The natural frequency for this spectrum is 26 Hz. And that of maximum Velocity is 21.7 mm/s.

Table 20: Maximum Velocity for single PGT, without and with phasing arrangement obtained from FFT analyzer

RPM	Maximum Velocity(mm/s)		
	Single PGT	With Phasing	Without Phasing
100	6.6	5.8	6.8
200	5.7	3.1	7.1
300	8.3	7.2	8.4
400	15.9	15.8	15.95
500	21.5	17.3	21.7

Summary

This velocity spectrums gives the peak value of velocity for single PGT, phasing and without phasing arrangement for various speeds of motor. This conclude that the velocity of phasing arrangement is reduces as compared to without phasing & single PGT arrangement for the various speeds.

CHAPTER 6

CONCLUSION

The experimental setup is developed for measuring the vibration and noise in planetary gear train. From the readings and spectrum obtained from the FFT analyzer and noise measuring instrument the vibration and noise are measured. The Vibration and Noise are measured by phasing and without phasing arrangement. After comparing the results obtained from FFT analyzer and sound level meter it is noticed that the vibrations are reduced up to 10 % and noise is reduced up to 8%. This shows that phasing of gears helps to reduce the vibration and noise in PGT.

FUTURE SCOPE

The vibration and noise in planetary gear train is reduced by phasing arrangement between two gear pair. This method is only limited for planetary gear train having fixed number of teeth. The phasing angle is depend on the number of teeth of gear pair. This method is helpful for automobile sector for reducing the vibration in gear train. By using phasing of gear the vibration in the conventional metal gears like cast iron, steel, Aluminum etc. may reduce which will the quality of system.

REFERENCES

[1] S. H. Gawande, S. N. Shaikh, R. N. Yerrawar, and K. A. Mahajan, "Noise level reduction in planetary gear set," Journal of Mechanical Design & Vibration, vol.
2, no. 3, pp. 60–62, 2014.

[2] D. L. Seager, "Conditions for the neutralization of excitation by the teeth in Epicyclic Gearing," Journal of Mechanical Engineering Science, vol. 17, no. 5, pp. 293–298, 1975.

[3] W. E. Palmer and R. R. Fuehrer, "Noise control in planetary transmissions," SAE Technical Paper 770561, 1977.

[4] A. Kaharamam and C. Yuksel, "Dynamic tooth loads of planetary gear sets having tooth profile wear," Mechanism and Machine Theory 39 (2004) 695–715. Science direct (2004)

[5] R.G. Parker, "Physical explanation for the effectiveness of planet phasing to suppress planetary gear vibration," Journal of Sound and Vibration, vol. 236, no. 4, pp. 561–573, 2000.

[6] C. Gill-Jeong, "Numerical study on reducing the vibration of spur gear pairs with phasing," Journal of Sound and Vibration, vol. 329, no. 19, pp. 3915–3927, 2010.

[7] Y. Chen and A. Ishibashi, "Investigation of noise and vibration of planetary gear drives," Gear Technology, vol. 23, no. 1, pp. 48–55, 2006.

[8] R. G. Parker, V. Agashe, and S.M. Vijayakar, "Dynamic response of a planetary gear system using a finite element/contact mechanics model," Journal of Mechanical Design, Transactions of the ASME, vol. 122, no. 3, pp. 304–310, 2000.

[9] P. Velex and L. Flamand, "Dynamic response of planetary trains to mesh parametric excitations," Journal of Mechanical Design, vol. 118, no. 1, pp. 7–14, 1996.

[10] R. J. Drago, "How to design quiet transmissions," Machine Design, vol. 52, no. 28, pp. 175–181, 1980.

[11] R.G.Parker, X. Wu "Unique symmetry phenomena in the vibration of planetary gears" 200240, 2002

[12] W. Cheng, "Simulation of the stochastic vibration of spur gears," in Proceedings of the 16th International Conference on Computer Aided Production Engineering, Edinburgh, UK, November 1988.

[13] A.Kahraman and R. Singh, "Non-linear dynamics of a spur gear pair," Journal of Sound and Vibration, vol. 142, no. 1, pp. 49–75 1990.

[14] B. Torby, Spur-Gear Dynamics, vol. 13 of TRITA-MMK, Royal Institute of Technology, Stockholm, Sweden, 1995.

[15] A.Palermo, D. Mundo A.S., Lentini, R. Hadjit, P. Mas, W. Desmet "Gear noise evaluation through multi body TE-based simulations," Proceedings of Isma2010 Including Usd2010.

[16] K. Ariga, T. Abe, Y. Yokoyama, and Y. Enomoto, "Reduction of transaxle gear noise by gear train modification," SAE Technical Paper 922108, 1992.

[17] M. G. Donley, T. C. Lim, and G. C. Steyer, "Dynamic analysis of automotive gearing systems," SAE Technical Paper 920762, 1992.

[18] Yichao Guo "Analytical Study On Compound Planetary Gear Dynamics" (P.H.D. Thesis) The Ohio State University (2011)

[19] W. Hellinger, H. Raffel, and G. Rainer, "Numerical methods to calculate gear transmission noise," SAE Technical Paper 971965, 1965.

[20] I. Nurhadi, "Investigation of the influence of gear system parameters on noise generation " [Ph.D. thesis], The University of Wisconsin, Madison,Wis, USA, 1985.

[21] Majid Meharabi, Dr. V. P. Singh " Vibration Analysis of Planetary Gear System " (IJARME) Vol.3 Iss-2, 2013.

[22] P. J. Sweeney, Transmission error measurement and analysis [Ph.D. thesis], University of New South Wales, New South Wales, Australia, 1995.

[23] A. H. Middelton, "Noise testing of gear boxes and transmissions using low cost digital analysis and control techniques," SAE Technical Paper 861284, 1986.

[24] A Al-shyyab, K Alwidyan, A Jawarneh, HTlilan "Non-linear dynamic behaviour of compound planetary gear trains: model formulation and semi-analytical solution" JMBD197. 223 Part K (2009).

[25] F. B. Oswald, D. P. Townsend, M. J. Valco, R. H. Spencer, R. J. Drago, and J.W.Lenski Jr., "Influence of gear design on gearbox radiated noise," Gear Technology, vol. 15, no. 1, pp. 10–15, 1998.

[26] R. G. Schlegel and K. C. Mard, "Transmission noise control approaches in helicopter design," in Proceedings of the ASME Design Engineering Conference, ASME paper 67-DE-58, New York, NY, USA, 1967.